Backcountry Press Presents

Field Guide to Manzanitas

California, North America, and Mexico

Michael Kauffmann
Tom Parker
Michael Vasey

Photographs by Jeff Bisbee

A. patula in the Marble Mountain Wilderness.

Order this book online at www.backcountrypress.com

Second Edition - first printing

All text by authors, all photos by Jeff Bisbee unless otherwise noted
Book layout and maps by Backcountry Press
Front cover: *Arctostaphylos edmundsii* along the central California Coast.

Published by Backcountry Press | Kneeland, California

ISBN 978-1-941624-15-9
Library of Congress control number 2015902037

MIX
Paper from responsible sources
FSC® C005010
FSC www.fsc.org

BACKCOUNTRY PRESS
Humboldt County, CA

Table of Contents

Discussion

8 Preface

9 Quick Guide

10 Introduction to Manzanitas

11 • The California Floristic Province

13 • Origin, Ecology, and Diversification in *Arctostaphylos*

19 • Conservation Implications

19 • Beyond the California Floristic Province

Regional Keys to Species and Subspecies

22 Manzanita Characteristics

Regional Keys:

28 • Klamath Mountain

30 • San Francisco Bay Region

33 • Central Coast Region

37 • Sierra Nevada Region

38 • Southern California Region

39 • Baja California Region

41 • Outside the California Floristic Province

Manzanita Treatments

43 *A. andersonii* - Santa Cruz Manzanita

44 *A. auriculata* - Mount Diablo Manzanita

45 *A. australis* - Baja Manzanita

A. bakeri

46 subsp. *bakeri* - Baker's Manzanita

47 subsp. *sublaevis* - The Cedars Manzanita

48 *A. bolensis* - Bola Manzanita

A. canescens

49 subsp. *canescens* - Hoary Manzanita

51 subsp. *sonomensis* - Sonoma Canescent Manzanita

52 *A. catalinae* - Catalina Island Manzanita

53 *A. columbiana* - Columbia Manzanita

54 *A. confertifolia* - Santa Rosa Island Manzanita

55 *A. crustacea* subsp. *crustacea*
subsp. *crinita* - Crinite Manzanita
57 subsp. *eastwoodiana* - Eastwood's Brittleleaf Manzanita
subsp. *rosei* - Rose's Manzanita
subsp. *insulicola* - Island Manzanita
subsp. *subcordata* - Santa Cruz Island Manzanita
58 *A. cruzensis* - Arroyo De La Cruz Manzanita
59 *A. densiflora* - Vine Hill Manzanita
60 *A. edmundsii* - Little Sur Manzanita
62 *A. franciscana* - Franciscan Manzanita
63 *A. gabilanensis* - Gabilan Manzanita
64 *A. glandulosa* subsp. *glandulosa* - Eastwood's Manzanita
64 subsp. *adamsii* - Adam's Manzanita
subsp. *atumescens* - Punta Banda Manzanita
65 subsp. *crassifolia* - Del Mar Manzanita
subsp. *cushingiana* - Cushing's Manzanita
subsp. *erecta* - Pedregoso's Manzanita
subsp. *gabrielensis* - San Gabriel Manzanita
66 subsp. *howellii* - Howell's Manzanita
subsp. *leucophylla* - Whiteleafed Manzanita
subsp. *mollis* - Transverse Range Manzanita
68 *A. glauca* - Bigberry Manzanita
69 *A. glutinosa* - Schreiber's Manzanita
70 *A. hispidula* - Gasquet Manzanita
A. hookeri
72 subsp. *hearstiorum* - Hearst's Manzanita
73 subsp. *hookeri* - Hooker's Manzanita
74 *A. hooveri* - Hoover Manzanita
75 *A. imbricata* - San Bruno Mountain Manzanita
76 *A. incognita* - Incognito Manzanita
77 *A. insularis* - Santa Cruz Island Manzanita
78 *A. klamathenis* - Klamath Manzanita
79 *A. luciana* - Santa Lucia Manzanita
78 *A. malloryi* - Mallory's Manzanita

A. manzanita subsp. *manzanita* - Common Manzanita
81 subsp. *elegans* - Konocti Manzanita
subsp. *glaucescens* - Glaucous Manzanita
subsp. *laevigata* - Contra Costa Manzanita
82 subsp. *roofii* - Roof's Manzanita
subsp. *wieslanderi* - Wieslander's Manzanita
A. mewukka
84 subsp. *mewukka* - Indian Manzanita
85 subsp. *truei* - True's Manzanita
A. montana
86 subsp. *montana* - Mount Tamalpais Manzanita
87 subsp. *ravenii* - Presidio Manzanita
88 *A. montaraensis* - Montara Mountain Manzanita
89 *A. montereyensis* - Monterey Manzanita
90 *A. moranii* - Moran's Manzanita
91 *A. morroensis* - Morro Manzanita
92 *A. myrtifolia* - Ione Manzanita
A. nevadensis
93 subsp. *nevadensis* - Pine Mat Manzanita
94 subsp. *knightii* - Knight's Pine Mat Manzanita
95 *A. nissenana* - El Dorado Manzanita
96 *A. nortensis* - Del Norte Manzanita
A. nummularia
98 subsp. *nummularia* - Glossyleaf Manzanita s
99 ubsp. *mendocinoensis* - Pygmy Manzanita
100 *A. obispoensis* - Bishop Manzanita
101 *A. ohloneana* - Ohlone Manzanita
102 *A. osoensis* - Oso Manzanita
104 *A. otayensis* - Otay Manzanita
105 *A. pacifica* - Pacific Manzanita
106 *A. pajaroensis* - Pajaro Manzanita
107 *A. pallida* - Alameda Manzanita
A. parryana
108 subsp. *parryana* - Parry's Manzanita
109 subsp. *desertica* - Desert Manzanita
subsp. *tumescens* - Interior Manzanita

A. patula

110 subsp. *patula* - Greenleaf Manzanita s

111 ubsp. *gankinii* - Gankin's manzanita

112 *A. pechoensis* - Pecho Manzanita

A. peninsularis

114 subsp. *peninsularis* - Peninsula Manzanita

115 subsp. *juarzensis* - Sierra Juarez Manzanita

116 *A. pilosula* - Santa Margarita Manzanita

A. pringlei

118 subsp. *pringlei* - Pinkbract Manzanita

119 subsp. *drupacea* - California Pinkbract Manzanita

120 *A. pumila* - Sandmat Manzanita

121 *A. pungens* - Mexican Manzanita

A. purissima

122 subsp. *purissima* - La Purisima Manzanita

123 subsp. *globosa* - Globose Manzanita

124 *A. rainbowensis* - Rainbow Manzanita

125 *A. refugioensis* - Refugio Manzanita

126 *A. regismontana* - King's Mountain Manzanita

127 *A. rudis* - Sand Mesa Manzanita

128 *A. sensitiva* - Coinleaf Manzanita

129 *A. silvicola* - Bonny Doon Manzanita

A. stanfordiana

130 subsp. *stanfordiana* - Stanford's Manzanita

131 subsp. *decumbens* - Rincon Ridge Manzanita s
subsp. *raichei* - Raiche's Manzanita

A. tomentosa

132 subsp. *tomentosa* - Woolyleaf Manzanita

133 subsp. *bracteosa* - Glandular Woolyleaf Manzanita
subsp. *daciticola* - Dacite Manzanita
subsp. *hebeclada* - Explorer's Manzanita

A. uva-ursi

134 subsp. *uva-ursi* - Bearberry or Kinnikinnick

135 subsp. *cratericola* - Guatemala Bearberry

136 *A. virgata* - Marin Manzanita

137 *A. viridissima* - White-haired Manzanita

A. viscida
138 subsp. *viscida* - Whiteleaf Manzanita
139 subsp. *mariposa* - Mariposa Manzanita
140 subsp. *pulchella* - Coast Whiteleaf Manzanita

Destinations

144 Northwest California
146 Northern Sierra Nevada
148 San Francisco Bay
150 Monterey Bay
152 Central California Coast
154 Southern California
156 Baja California

Tying a bow around it...

Manzanita Lists By Region
159 By California Counties
165 In Baja California
166 Outside California
167 Glossary
168 Suggested Reading
170 Manzanita Relatives
174 Photographers and Authors

Special Thanks:

Contributing reviewers, in alphabetical order:
Jeff Bisbee
Jeffrey Kane
Julie K. Nelson
Denise Newman
Allison Poklemba
Jacob Pounds
John O. Sawyer
David Rains Wallace

All photos by **Jeff Bisbee,** unless otherwise noted.

Botanical illustrations in the key section by **Gregory Arena**.

Preface

Plant and animal distribution has always intrigued me—I can read range maps for hours, hence my affinity for creating them—and I often choose destinations based on unique distributions of plants or animals. One such trip was a backpack to Mopah Springs in the Turtle Mountain Wilderness where I was in search of a grove of California fan palms. Many miles of desert scrub surround this remote oasis: the clustered palms drink the waters of a spring-fed pond. Its needs met, the species endures a climate that otherwise allows only cactus and hearty shrubs to survive.

Resting in the palms' shade, I pondered various hypotheses about their origins. Could Spanish prospectors have brought seeds hundreds of years ago? Did Native Americans plant them thousands of years ago? Are they relicts from the Pleistocene Epoch, a moister time ~12 thousand years ago when other palms perhaps inhabited nearby lake shores? Maybe one day genetic work will shed more light on the palms' past—but what joy to interpret a place by hypothesizing how ecological needs, and consequent distributions, set limits on the survival of flora and fauna.

In the California Floristic Province, the genus *Arctostaphylos* is a particularly fine illustration of how long-term dispersal events lead to colonization and consequent adaptive radiation. Fossil records show that this genus has been migrating and adapting to climatic shifts for at least 15 million years. However, only in the past few million years has *Arctostaphylos*, commonly called manzanita for its berries' resemblance to small apples, found its promised land. The California Floristic Province's exceptionally diverse range of habitats, particularly of ones that provide a taste of the suboptimal, is perfect for manzanitas. A synergistic mix of climate stability, soil variability, topographic volatility, and fire frequency provides an alignment of biotic and abiotic factors. Like many other California evergreens (including my beloved conifers) these hardy plants have benefited from inhospitable environments wherein competition from many plants is reduced and their own adaptability to poorer growing sites allows them to thrive. This, somewhat ironically, has made the unassuming "little apple" the most species-rich shrub genus in the California Floristic Province. You will learn much more of this compelling story from this book.

The second edition of *A Field Guide to Manzanitas* would not have been possible without the efforts of my distinguished collaborators. Photographer Jeff Bisbee is the consummate plant explorer—searching out manzanita species in the field, across growing seasons, and capturing them in every phase of their life cycle. The lifelong work of botanists Tom Parker and Michael Vasey, two of the world's manzanita experts, provided the discussion of the genus that follows, specifically crafted regional keys, species treatments, and consequential edits and additions five years after the first edition was printed. This collaboration arose because I had a passion to learn more a about this beautiful group of plants. I am thankful that I have an updated guide to take with me on plant exploring adventures for years to come.

We invite you to explore, through the manzanitas, one extraordinarily rich aspect of our world-class oasis called California. Take this book for a walk or a ride and get to know these iconic plants whose ecology helps define our dynamic landscape.

– Michael Kauffmann

Quick Guide:

- *A Field Guide to Manzanitas* treats of all 107 taxa in the world. It is written for anyone wishing to better appreciate the natural world and learn more about the diversity of plants within the California Floristic Province.
- Each manzanita species is accorded images, range map, and description of characteristics. Treatments are organized alphabetically by species' Latin name. Taxonomy is dynamic, particularly with this genus; therefore, scientific names may change or new species may be described. Updates will be reflected in future printings of this book and updated on the book's website.
- If you wish to seek out specific locations and find a particular species, refer to the **Observations** section in the treatments for tips.
- The **Destinations** section suggests 28 sites over 7 regions of the California Floristic Province that offer exceptional manzanita explorations.
- Lists by California counties, beginning on pg. 159 will also help with identification based on your location.
- Though we tried to avoid the use of language that was too technical, sometimes we had to use botanical terms. Refer to the **Glossary** on pg. 168.
- **Suggested Reading** on pg. 164 will direct readers to more resources.
- Visit the book's website (**www.backcountrypress.com/manzanitas**) for more information, to share a discovery, or make a suggestion about the book.

Comments on Range Maps

The range maps have been created from a collection of sources, but most were assimilated using direct observations from citizens and scientists who have taken the time to report them via collections for herbaria. These collections are then either digitized by herbaria or directly reported by these naturalists to databases like CalFlora, the Oregon Flora Project, or the Consortium of Pacific Northwest Herbaria. These data were analyzed and filtered to what you see herein.

While we believe these maps are highly accurate, species could be found outside the shaded regions. We hope the maps are helpful for identifying this difficult genus but all species have outliers and new discoveries are always being made. For some species, we have included labels on the maps to help identify individual populations. For the more common, or widespread manzanitas, county lines, mountain ranges, rivers, national forest boundaries, or other large-scale landmarks included on the maps are the best way to track down specific locations. These maps will be modified in the future as new data is submitted by scientists, or habitat is lost due to fire events, urban sprawl, or climate change. Please contact us with questions or share your discoveries on iNaturalist.

Regional Keys

Great effort was put into the creation of keys for manzanita species by region. While most of the regions are within California, for the first time there is a key (as well as treatments) for manzanita species in northern Baja California. There is also a key to all species outside the California Floristic Province.

Introduction to Manzanitas

Manzanitas are the "rock stars" of woody shrub diversity in California. Ranging from the Sierra Nevada mountains to the coastal bluffs along the Pacific, from temperate rainforests along the north coast to arid mountain slopes in Southern California, a wealth of manzanita species and subspecies can be found in an astonishing array of environments. Manzanitas occur on serpentines, dunes, volcanic soils, sandstone outcrops, dense shale, granite, gabbro–the list goes on.

Even with this astounding species and habitat diversity, there are common denominators that characterize this group. Most manzanitas depend upon fire for their regeneration. This is the key to why manzanitas are so prominent in the Mediterranean-type climate of California. Here, high-severity (in addition to mixed- and low-severity) wildfires have shaped the region during its long summer dry season. Another common feature of most manzanitas is that they occur, and often dominate, California's evergreen fire-adapted shrubland known as chaparral. Similar shrublands are found in only four other regions on the planet (see next section for more information). Finally, while manzanitas are found on a wide variety of substrates, preferred soils are typically shallow, rocky, and/or nutrient poor.

The vast majority of *Arctostaphylos* taxa are found within the California Floristic Province (CFP), an area that encompasses southwestern Oregon, much of California (excluding the deserts), and northern Baja California (fig. 1). This area effectively represents the Mediterranean-type climate region in North America and Mexico wherein 106/107 (99%) of *Arctostaphylos* species and subspecies are found with only seven of these taxa (6%) (*A. uva-ursi* subsp. *uva-ursi* & subsp. *cratericola, A. nevadensis, A. columbiana, A. patula* subsp. *patula, A. pungens, A. pringlei* subsp. *pringlei*) extending beyond the CFP. Of these 107 taxa, 63 (60%) are considered local endemics; i.e., taxa whose distribution covers less than 1000 km^2 (an area less than 20 miles by 20 miles). Further, of these 63 local endemic taxa, 50 (82%) are found in near-proximity to the Pacific Coast. This richness of local endemics, particularly along the California coast, makes manzanitas exciting to observe, challenging to identify, and important from a conservation perspective.

Our purpose in this field guide is to "lift the veil" on this quintessentially Californian shrub genus so that professional and amateur botanists alike are able to more fully appreciate the beauty, complexity, diversity, distinctive habitats, and conservation value of wild manzanita populations. We rely largely on the magnificent photographs of Jeff Bisbee to introduce this topic. Jeff has spent countless hours documenting each species with photos from various life stages. His work is augmented with information about where these taxa occur and how to identify them.

We have also prepared special regional keys to help guide readers in place-based field identification in California, across North America, and into Mexico. The two most recent treatments of *Arctostaphylos* can be found in Volume 8 in *The Flora of North America* (2008) and the *Jepson Manual, 2nd Edition* (2012), as well as online (e.g. the Jepson Interchange). Each has a single technical key that is challenging to use unless you are a trained botanist (and sometime not even then). Also, neither of these treatments includes manzanitas from Baja California. Therefore, this field guide provides the most up-to-date treatments of, and keys for, manzanita taxa that are endemic to Baja.

While the two treatments mentioned above provide the best technical descriptions of *Arctostaphylos* taxa in North America and California, we have intentionally tried to minimize technical terms in our regional keys to give them a more universal appeal. We used the regional approach because most manzanita taxa are so restricted in their distribution that literally knowing where you are can be the best step toward proper identification. Nonetheless, the reader will discover that keys for the San Francisco Bay region and Central California Coast (including the northern Channel Islands) are still complex and contain a great number of taxa. We encourage our noble readers to become familiar with the treatments and to use them to help validate identifications made on the basis of this field guide.

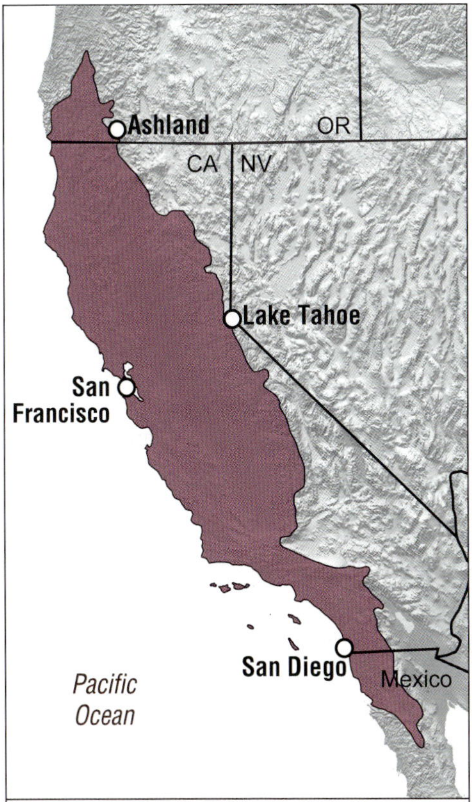

Figure 1: California Floristic Province

We also should point out that there remain variants, not yet described, of some of these species. In other cases some areas–like in Baja California, Mexico–have not been well explored and botanized. Keep in mind there are always outliers from known ranges. Thus, your own exploring and documentation is a wonderful way to develop an appreciation for variation and the unusual diversity of habitats in which they are found, and to fully appreciate the great diversity of the California Floristic Province.

The California Floristic Province

The story of the California Floristic Province is one that has developed across millions of years of tectonic activity, climatic changes, botanical migrations, and subsequent evolutionary events. Today, what makes the region a botanical hotspot is not only the variety of environments offered to plants—from beaches to mountain summits—but also the complex climates and array of substrates which nurture a collection of microsites for plants to inhabit. While numerous plant genera could be used to represent endemism and evolution within the CFP, manzanitas might just be the finest of all examples. Manzanitas have a nearly-exclusive relationship with the CFP and have evolved to fit the regional Mediterranean-type climate, especially during the ice ages that occurred during the past 2.5 million years. For that reason, let's explore how the regional climate, and consequent flora, came to be.

Mediterranean-type climates are rare, today occurring in only 5 areas around the world (fig. 2). These places share several characteristics in common, despite the vast distances between them. These attributes include hot dry summers and cool wet winters, which when combined, favor shrublands as a dominant vegetation type. These shrublands tend to burn frequently and intensely–often as stand-replacing events. Mediterranean regions also share an unusually high number of locally endemic plants when compared to other climate types at similar latitudes.

Many of these local endemic species occur on harsh, nutrient-poor soils. Within the Mediterranean-type climate regions, there are local gradients of water availability and temperature extremes and these factors are also associated with patterns of plant species diversity. In summary, recent hypotheses concerning the diversity of regional Mediterranean-type climate floras is that they are forged by the synergistic interaction of geology, climate, and fire together rather than being attributed to one or the other of these factors.

The history of vascular plants in the California Floristic Province is summarized by Dr. Constance Millar in the 2nd Edition of the *Jepson Manual*'s flora of California. Much of the early fossil evidence for California's drought-adapted woody vegetation is found in the western Great Basin where there was once a high regional plateau (the Nevadaplano) near the eastern border of the ancient Sierra Nevada. This high plateau was pulled apart and collapsed, eventually resulting in the current Basin and Range Province. The earliest manzanita fossils are found in western Nevada about 15.2 million years ago (mya). This is a time that followed a climatic optimum (warming period) during the Middle Miocene (17-15 mya) that would have favored woody evergreen genera like manzanitas.

Present-day California also experienced considerable geological disruptions including volcanic activity, mountain range uplifting, and a cooling and drying trend

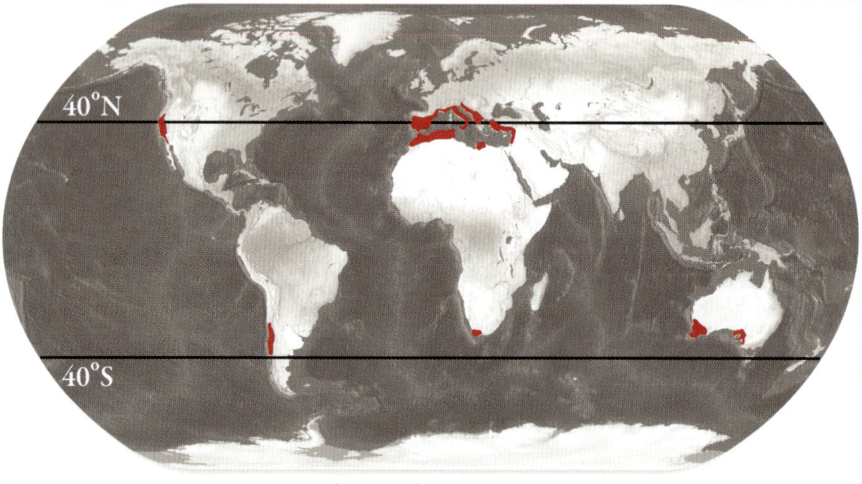

Figure 2 Areas with a Mediterranean-type Climate

following the Middle Miocene when moisture-dependent species would have migrated towards the coast. Around 7 mya, the California Current[1] became active and it appears that by 5 mya a seasonal Mediterranean-type climate was regionally established. Following this time, the region gradually moved into a series of Ice Ages in the Pleistocene, which began about 2.6 mya. Roughly 40 glacial and inter-glacial phases have taken place since then and each of these climatic oscillations occurred in the context of California's extremely long summer dry season resulting in dependable fire cycles, even during glacial phases. Over the millennia, particularly in unglaciated regions such as in the coast range, plants migrated back and forth in an opportunistic shuffle, with dominant species in fire-prone habitats shifting from one place to another tracking climate shifts, and moving into and out of unusual soil conditions and slope orientations.

Many evergreen shrubby genera, such as *Adenostoma, Ceanothus, Heteromeles, Garrya, Pickeringia, Lyanothamnus, Carpenteria,* and *Arctostaphylos* would have become widespread over the past five million years as shrubland vegetation spread due to increased wildfires once the Mediterranean-type climate became established. Interestingly, other than *Arctostaphylos* and *Ceanothus,* these genera did not radiate into numerous local endemic species even though they arise from equally venerable lineages that date back to the days of the Nevadaplano, millions years ago. The big difference between *Arctostaphylos* and *Ceanothus* compared to these other genera is their response to fire (many are obligate seeders[2]) as well as a genetic make-up that favors adaptability to an evolving landscape. To understand this distinction, we turn to a discussion about the origin, ecology, and diversification of *Arctostaphylos.*

Origin, Ecology and Diversification

Manzanitas are members of the heath family (Ericaceae), an ancient lineage of woody plants that dates back to the late Cretaceous (~ 110 mya). The heath family has a global distribution with members prominent on all continents in both hemispheres, with the exception of Antarctica. Well known plants include rhododendrons, cranberries, huckleberries, and blueberries. Despite a wide distribution and great diversity, members of this family (like *Arctostaphylos*) have important characteristics in common. They typically occur in fringe environments, preferring soils that are either nutrient-poor, waterlogged, arid, or in extremely cold environments. Adaptation to these harsh conditions has partly been tied to a symbiosis with fungi that produce mycorrhizae, or literally "fungal roots." These fungi penetrate the roots of shrubs in the heath family and essentially trade nutrients and water they obtain from the surrounding environment for carbon energy fixed by woody plants through photosynthesis. This mutualistic relationship is one key to the success of members of the heath family in these suboptimal environments.

---

1. *A current in the Pacific Ocean that moves southward along the western coast of North America, beginning off southern British Columbia, and ending off southern Baja California. This pattern keeps the coastal waters cool and produces characteristic coastal fog and cool summer temperatures immediately adjacent to the ocean.*

2. *Plants that are killed by fire, but have large, fire-activated seed banks that germinate, grow, and mature rapidly following a fire, in order to reproduce and renew the seed bank before the next fire.*

Manzanitas belong to a relatively small and distinct subfamily within the Ericaceae called the Arbutoideae (the madrone subfamily). There are only six genera recognized within the madrone subfamily: *Arbutus, Arctous, Arctostaphylos, Comarostaphylis, Ornithostaphylos,* and *Xylococcus. Arbutus* is familiar to many in western North America as the madrone, with other species occurring in Arizona, Texas, Mexico, and the Mediterranean region of Europe and north Africa. The strawberry tree, *Arbutus unedo*, is a common horticultural plant in our region but a native of the western Mediterranean. *Arctous* is unique in this subfamily in that it is deciduous and only occurs in high elevations or at high latitudes in the northern hemisphere. Historically, this genus has been treated as a manzanita. Other members of this subfamily are centered in western North America and adjacent Mexico with relatively few extending beyond this region. *Comarostaphylis* (summer holly) has a number of species in Mexico, with one species that is found in southern California and the Channel Islands. *Xylococcus* (mission manzanita) and *Ornithostaphylos* (Baja birdbush) have only one species (monotypic) each and occur in northern Baja into southern California. See the Manzanita Relatives section on page 170 to learn more.

Based upon its relationship to primitive mycorrhizal fungi, the madrone subfamily may represent one of the most ancient lineages within the heath family. While fossil, distributional, and molecular evidence suggests that *Arbutus* is the oldest genus in this group (North American fossils exceed 50 mya), other genera such as *Comarostaphylis, Xylococcus,* and *Ornithostaphylos* appear to have arisen prior to *Arctostaphylos* and, possibly, *Arctous*. Based on molecular genetic evidence, these latter five genera are more closely related to one another than to *Arbutus*. This separate lineage has a number of traits that have changed from those found in *Arbutus*. One important trait is how the fruit have evolved—from a succulent berry in *Arbutus* to berry-like drupes in the other five genera. Madrones have small, unprotected seeds while the other genera have seeds surrounded by a hard, stony layer (endocarp), called stones or nutlets. Sometimes these stones are separated within the fruit and sometimes fused into one solid stone[3]. While birds cannot benefit much from these stony seeds, animals like rodents are able to crack them and consume the nutritious embryos. The one exception to this situation may be *A. uva-ursi* which has an unusual amount of mealy mesocarp between its bright red exocarp and small, stony seeds. The wide distribution of *A. uva-ursi* may reflect bird dispersal of these unusually fleshy fruits within *Arctostaphylos*.

Through much of the last several million years, manzanita taxa were present in what is now California as more favorable maritime climatic conditions "pulled" them towards the Pacific Coast. Ultimately, based on molecular evidence, we suspect that several relatively old lineages within *Arctostaphylos* became established near the coast, taking advantage of these less extreme climates. Many of these lineages are still present today in coastal and mountain refugia. However, other lineages seem to have gone through a relatively recent radiation and given rise to new manzanita taxa. To understand the processes that likely facilitated the rise of relatively recent taxa, we need to first consider the ecology of *Arctostaphylos*.

---

3. *Note: this is an important character when using the key.*

Ecology

Manzanitas tend to occur in sites with poorly developed, usually low nutrient soils. Mycorrhizal fungi are a principal reason manzanitas can tolerate these low nutrient conditions. One example physiologically is that their hard, evergreen leaves require fewer nutrients in their creation than other plants. In addition, these leaves decompose slowly and the high amount of organic acids in the leaves tends to acidify the soil when they finally do, which continues to maintain low soil nutrient levels. The fungal mutualists assist in breaking down the litter and cycling the nutrients back to the manzanitas through the fungal hyphae (thread-like structures). While most plants cannot tolerate these types of conditions, a number of evergreen species, like conifers, are similar in nutrient requirement and create similar conditions on nutrient poor sites. Consequently, plants with analogous requirements and tolerances tend to co-occur and it is common to find manzanitas associating with conifers and other evergreen plants. In fact, roots of manzanitas and several diverse plant species are connected by an extensive mycorrhizal network below ground that is vital to patterns of community succession over time.

Most manzanitas occur in areas with strong climatic seasonality. In California, summers are warm and generally rainless, creating dry and potentially stressful conditions for plants. Manzanitas express a number of ways of dealing with summer drought, including hidden traits in the wood (more below) or the orientation of their leaves. You may have noticed that most manzanitas hold their leaves nearly vertically; researchers have speculated that one function of this is to assist in lowering leaf temperatures during the mid-day heat, especially in summer, allowing plants to use less water to cool leaves by transpiration. This trait is not found in any other genus in the Madrone subfamily and has also resulted in a division of leaf types within the manzanitas. Some species have leaves held horizontally and have small openings (called stomata) in the leaves for carbon and water vapor exchange only on the lower leaf surface, while others hold their leaves upwards and have stomata on both sides. You will see these stomatal characters appear in the keys that follow.

While climate and soils are significant influences on manzanita ecology, most terrestrial systems also experience some type of fire regime. A fire regime is the overall pattern of fires experienced by vegetation over time–including intensity, seasonality, frequency, and size. Many shrub-dominated ecosystems in California's Mediterranean-type climate exhibit high severity stand-replacing fire regimes. This fire regime is characterized by fires of significant sizes; such large disturbances mean that the vegetation would have a difficult time dispersing back into the post-fire site from nearby unburned areas. Instead, chaparral species have adaptations to persist after stand-replacing wildfires. Because fire return intervals and other dimensions of the fire regime are relatively predictable, repeated fires lead to most plants exhibiting traits adaptive to fire. Different fire regimes will favor particular sets of fire-adaptive traits. The stronger the selection for fire-adaptive traits, the less the vegetation composition seems to change post-fire.

Vegetation types adapted to stand-replacing fires, however, do experience considerable change in post-fire environmental conditions. With the removal of the shrub canopy and most above-ground biomass, light levels become drastically higher and the potential for water deficit or temperature extremes increases, especially in arid locations. Seedlings are released from competition for light energy, but experience a harsh environment until they become well established.

Manzanitas have seeds that are dispersed in a dormant state and require, among other conditions, chemicals found in smoke to break that dormancy. Manzanitas are the only genus in the madrone subfamily known to have evolved this trait and this may have contributed to their success and radiation in fire-adapted habitats. Because of seed dormancy, manzanita seeds tend to accumulate in soil seed banks. If you recall, the evolutionary changes in fruits and seeds from *Arbutus* to *Arctostaphylos* hypothesizes that the fruits have switched from birds as dispersal agents toward mammals, like rodents. This shift was critical, because rodents tend to bury the seed to store it and hide it from other rodents. In this mutualistic relationship, rodents bury the seed deep enough so that the soil protects them from the high temperatures of wildfires.

The surviving seed are stimulated by fire and germinate in the next growing season—which, with fall rains, can occur in only a few months. The fires and the post-fire environments differ in conditions depending on local climates and soils and the types of seed and seed-eating mammals found in the region. Together, this accounts for the variability in fruit size and degree of fusion in seeds and stones that manzanitas display. Consider not only the characteristics as a way of determining species, but also why those variations may be important in particular places. You'll find that we use fruit and stone characteristics in many of the regional keys below.

Manzanitas have many other fire-adaptive traits. Dormant buds in woody structures allow resprouting of surviving individuals. When present, these structures are often prominent and are called burls (also ligno tubers). About one-third of manzanita species have burls, but they are commonly found in a number of different shrub genera in chaparral stands. As manzanita species with burls experience a sequence of fires and resprouting, they tend to form circles of plants. You may be able to spot some of these circles if the conditions are right. The centers of larger resprouting circles may have higher concentrations of organic matter, accumulated leaves and the slowly rotting remnants of former burls; high intensity fires can consume those old portions and sometimes burn the newer burls out from the inside. Look for such patterns when you visit sites that have recently burned (but be careful to not fall into the holes left behind by ashed burls!). Because burls contain high concentrations of dormant buds, mature plants that have burned before usually show multiple stems at the base. That is sometimes a way of determining if a burl is present. In any case, because about one-third of the species have burls and the other two-thirds do not, you'll find this characteristic is important in our keys.

One-third of manzanita species are facultative-seeders that regenerate post-fire by both seed and resprouting via burls. The remainder are obligate seeders that lose most or all of their entire adult population in a fire and depend on a seed bank for regeneration. To understand why obligate seeding is the current model in manzanita evolution, consider the climatic dynamics over thousands, or tens of thousands of years or more. In the case of the resprouting species, particular individuals can live for centuries, resprouting over and over, cloning new individuals as the burls expand with each fire cycle. But in that population, the rate of genetic change is limited, because most individuals live a long time by way of asexual reproduction. This suggests that populations may be unable to respond to rapid climatic changes that might occur in only hundreds of years. The obligate seeders, on the other hand, lose most adults in stand-replacing fires and new post-fire generations have to establish from more genetically-diverse seeds. Those populations consequently have greater flexibility to shift and adjust as circumstances require; traits that might have been rare and less important in older generations can emerge through natural selection and become critical in the newer generations within the lifetime of resprouting manzanitas.

Chaparral in the CFP is adapted to fire regimes with intervals of 30-150+ years. Within these time frames, chaparral returns to pre-fire conditions dominated by manzanitas and other evergreen shrubs because of the low mortality of resprouting individuals and the rapid reestablishment of shrub seedlings from fire-dependent seed banks. Critically, however, chaparral is adapted to a particular fire regime of a range of frequency, intensity, and seasonal timing so that excluding fire or increasing the frequency of fire reduces the long-term stability of this ecosystem type.

Diversification

Although fire regimes are one key to species diversity in manzanitas, it is not the only reason for their diversity. Because so many manzanitas are local endemics found along the California coast, it is instructive to consider what may be the driving forces shaping endemism in this region. Due to tectonic activity, shallow, rocky, nutrient-poor soils often occur as "edaphic islands" embedded within a matrix of deeper and more fertile soils. The coast also presents a steep climate gradient during the summer, ranging from cool and moist conditions at low elevations near the ocean to hot and dry conditions on inland ridges and slopes. Forest, coastal scrub, and grassland tend to dominate vegetation along the coast whereas chaparral is typically confined to the edaphic islands (fig. 3). Over time, especially during the Ice Ages, some have proposed that different manzanita species may have migrated onto and off of these soil islands during climatic oscillations.

Many species are capable of hybridization and, particularly within the obligate seeders, populations of hybrid origin may have been selected over repeated fire cycles resulting in new species (so-called neoendemics[4]). Because of these selective forces on localized sites, many species have not been able to spread widely across the landscape. On the other hand, formerly widespread manzanita species may now be confined to certain favorable sites where they are able to persist as relicts (so-called paleoendemics[5]). The rich number of localized endemics along the coast have arisen from a combination of both processes. Ultimately, similar to shrubland vegetation elsewhere in Mediterranean-type climate regions, diversification within *Arctostaphylos* may have taken place due to the long term interaction of fire, soils, and more favorable climate influences, particularly along the California coast.

4.Neoendemic manzanita species include, but are not limited to: *A. gabilanensis* & *A. refugioensis*

5. Paleoendemic manzanita species include, but are not limited to: *A. rudis* & *A. rainbowensis*

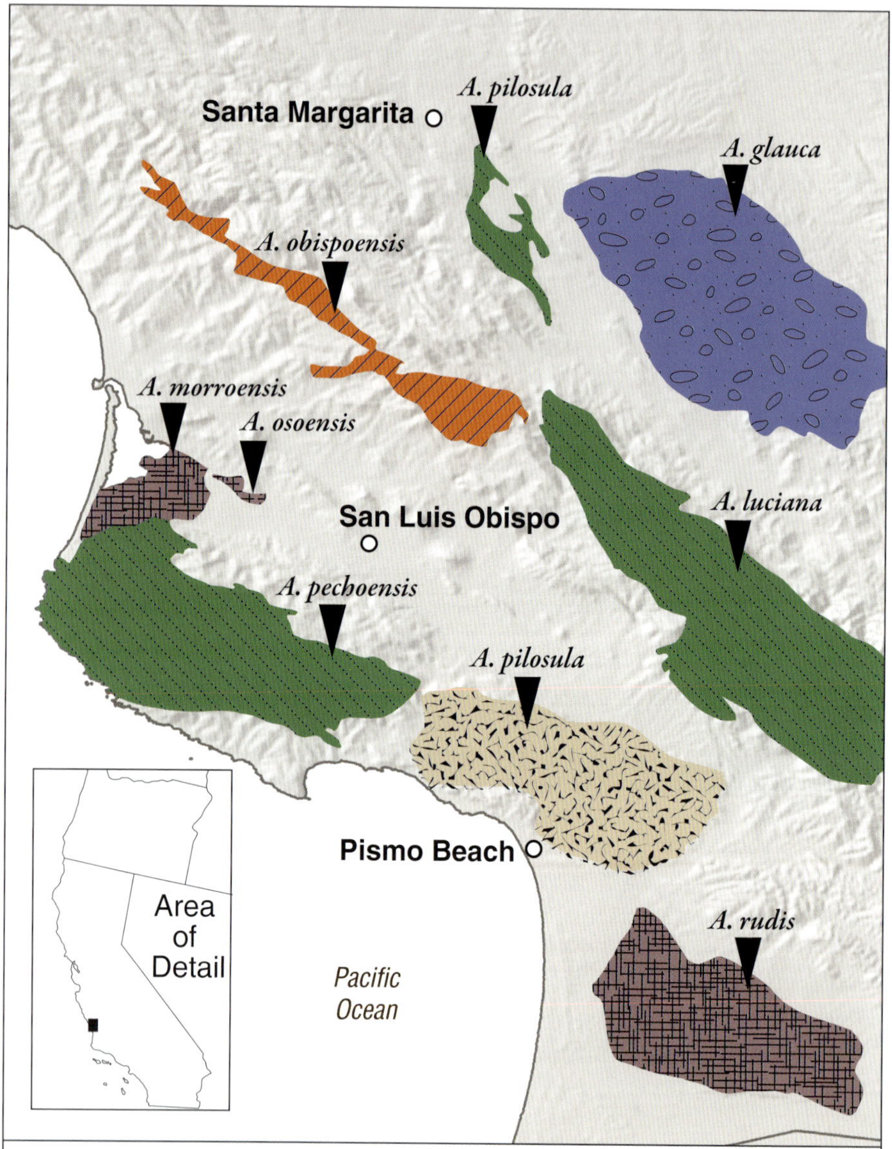

Examples of edaphic endemism in manzanitas* of San Luis Obispo County

Monterey Shale

Serpentine

Pleistocene Sands

Pismo Sandstone

Granite

**based on Wells (1980)*

Figure 3

Conservation Implications

In the early 1980s, interest emerged in conserving natural communities rather than species *per se*. Two natural communities in California that stood out as being both exceptionally rich in local endemics and at-risk due to human activities were Central Maritime Chaparral (Santa Barbara to San Francisco) and Northern Maritime Chaparral (from San Francisco to southern Oregon). Most endemic manzanita grow within maritime chaparral. Maritime chaparral is now protected under the California Coastal Act as 'Environmentally Sensitive Habitat Areas' and it is also given special consideration under the California Environmental Quality Act. Nonetheless, there are many challenges to the conservation of maritime chaparral and its host of local manzanita endemics.

An expanding human population in coastal California is putting increased development pressure on these islands of chaparral. As the urban-wildland interface expands, threats of fire impacting homes and structures adjacent to chaparral increase. With fire being an integral part of maritime chaparral ecosystems, vegetation is cleared around these areas to prevent fire impacts on human infrastructure. With suppression, natural fire regimes are disrupted which potentially impacts chaparral recruitment events. These factors are further influenced by climate change related phenomena. In particular, if the summer marine layer diminishes over time, manzanitas along the coast could well be replaced by more wide-spread drought tolerant species from the interior, such as chamise (*Adenostoma*) and bigberry manzanita.

Other anthropogenic threats include increased ignitions due to human activity which decrease the fire return interval, the use of mastication to manage for fire which can increase the number of non-native species which can then decrease the fire return interval. Short-term efforts should be made to curb development in chaparral areas and particularly coastal maritime chaparral with its high concentration of localized endemics. In the long-term, it may be necessary to establish local endemics in conservation gardens to await a time when suitable habitats materialize for relocation through assisted migration.

Beyond the California Floristic Province

While most manzanita species occur within the California Floristic Privince some species have extensive ranges outside this region. *Arctostaphylos uva-ursi* is the winner of the race around the globe, occurring throughout northern North America and across northern Eurasia. Not only is *A. uva-ursi* found in cold climates, such as on Greenland and Iceland, and alpine sites in Eurasia, it is also a common plant in coastal dune systems on both coasts of North America. *Arctostaphylos uva-ursi* subsp. *craterícol*a is a fascinating biogeographic oddity, likely disjunct from a previous Ice Age population, occupying "sky islands" in tropical Guatemala. Other exceptional taxa include *A. patula* subsp. *patula*, which occurs throughout most of the Rockies, and *A. pungens* that runs from Central California well into Mexico along the Sierra Madre.

The madrone subfamily has other mysteries as well. For example, *Arctous* is a total outlier for the subfamily as it is adapted to life in boreal and subartic regions. What might its relationships be to the other genera? The distribution of *Arctous* suggests that the madrone subfamily had at least one lineage that shifted away from dry climates to the very cold and short growing seasons of the far north (severe cold and drought often select for similar traits in plants). Another mystery to consider is that *Arbutus* is distributed in two sets of species, one set surrounding the Mediterranean Sea, and one set in Mexico and western North America. Molecular genetic data suggest that all the other genera (*Arctostaphylos, Comarostaphylis,* etc.) are more closely related to the Mediterranean species rather than the North American species of *Arbutus*, even though one would think that proximity might predict a closer relationship to the North America *Arbutus*. These distributional oddities indicate that the madrone subfamily has had an interesting history, which we are only just beginning to understand. This book will begin your exploration of *Arctostaphylos* within the California Floristic Province but there are other adventures awaiting beyond its regional borders.

The central coast range of California is the center of species diversity for Arctostaphylos. Pictured here is the Little Sur River.

Regional Keys

Manzanita Characteristics

Because it is a richly diverse genus, *Arctostaphylos* is challenging to identify at the species and subspecies levels. California alone has 95 recognized manzanita species and subspecies (taxa). However, there is some good news. In most cases, you will only have to face them one or a few at a time, depending on your location. In fact, knowing where you are is the first step to keying manzanitas successfully. This is because the majority of manzanita taxa are localized. For this reason, we have opted to create seven regional keys to simplify identification. Once in the field, look at the entire population before starting to examine individuals. You will probably be able to tell if there is more than one species in a given habitat by noting commonalities distributed throughout the localized population. Manzanita characteristics are relatively visible, and with a 10X hand lens, almost all the characteristics can be viewed in the field.

When attempting identification, become systematic in your approach to individual plants by developing an algorithm that works for you. For example, start by examining the base, move to the main stems, youngest stems, leaves, immature inflorescences, and then the fruit. Ironically, flowers rarely tend to be of much help in keying. We have structured the keys to focus on characters that are visible and available across as much of the year as possible, like leaves, immature inflorescences or fruit. Remember that field observations (or relevant photos) of burls, bark, and habitat are critical because these characters cannot be reconstructed from specimens once you are back home. Ideally, one should look at a number of individuals in the population rather than just one.

There are good reasons why manzanitas are difficult to key and why it is important to approach them as populations rather than individuals. As discussed in the introduction, *Arctostaphylos* is a venerable genus, dating back at least 15 million years to the middle Miocene. Most taxa (about two-thirds) are diploids (n=13 base chromosomes) but many others are tetraploids (n=26 base chromosomes). Based on molecular genetic analyses, we find that the diploids tend to be divided into two deep lineages. Species within these lineages do not hybridize readily with members of the other lineage, suggesting partial genetic incompatibility between the two. So when you find multiple species together, they will tend to be from different lineages or ploidy levels. A good example is Manzanita Park in Prunedale, California (see pg. 149), where you can find *A. hookeri* (diploid from one lineage), *A. pajaroensis* (diploid from another), and *A. crustacea* (tetraploid). Visit our website (manzanitas.backcountrypress.com) to find lists of species within these lineages.

Species within the deep lineages, by contrast, tend to be fertile with each other and are primarily isolated by ecological factors. Human disturbances or naturally occurring intermediate conditions, therefore, can pave the way for hybrids to successfully establish. For example, a famous site for hybrids is around Waldo, Oregon where mining activities have disturbed the soil. A variety of crosses between *A. viscida* (here preferring serpentine) and *A. canescens* (here preferring sandstone) were found on disturbed sites. By contrast, a stabilized hybrid (*A.* x *campbellae*) between *A. glandulosa* and *A. crustacea* occurs in a transitional zone between interior and coastal habitats in the Mount Hamilton region of Santa Clara County where neither typical parental species is present. In the Rocky Mountains, where moist subalpine conifer forests meet montane ponderosa pine forests bringing *A. uva-ursi* and *A. patula* together, the hybrid *A.* x *coloradensis* can be found intermixed with these populations. Unlike these relatively well known hybrids, where different manzanita species co-occur, it is not uncommon for some hybrid-like individuals to be found in the midst of their mixed populations, creating difficulty in keying individuals. This is why taking a population approach is the best policy when trying to identify to the species level.

In this field guide, we do not include hybrids or forms in the keys or descriptions, however, we do identify where these are known to occur in our table of distributions by county in California. For more information about these entities, the reader should consult local floras or Wells (2000), which presents a good taxonomic history on hybrids and general information on where they occur.

With these caveats in mind, below we identify the main key characters for manzanita taxonomy, covering stems, leaves, inflorescences, flowers, and fruits, that should lead the reader on an adventure exploring the fascinating diversity in this genus.

STEMS. About one-third of manzanita species will resprout after a fire from an enlarged woody structure called a **burl**. When first examining an unknown species, look to the base of the main stem for burls. One trick is to look uphill for these structures, not downhill where litter can cover them. If you see a number of variably-sized stems clustered together, that hints at a burl. These structures are not only critical for identification, but provide you with knowledge about a species ecology and evolution.

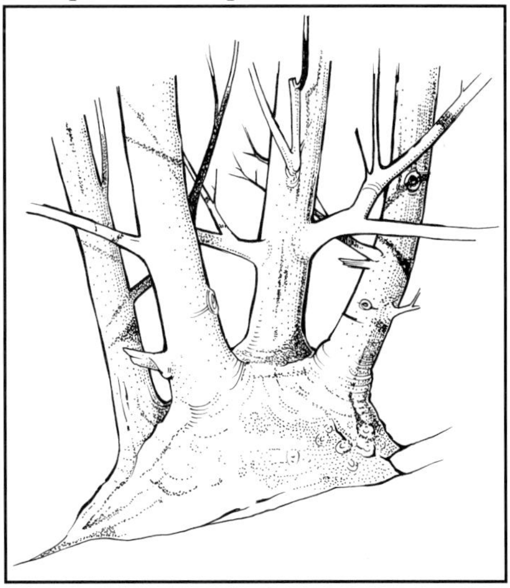

burl

Wildfires have been key for manzanita speciation and have caused a wide variety of life histories to arise. The burl is actually a swollen woody mass with a concentration of dormant buds from which plants can sprout after the upper canopy is killed by fire. As a consequence, resprouting individuals may be very old, but the above ground stems will only date as far back as the last fire. Manzanitas lacking a burl are killed by fire and recruitment of new populations after fire is completely dependent on the persistent soil seed banks (see introductory section for more). When you are examining the base of the stem for the presence of burls, you're actually asking a question about the ecology and evolution of manzanitas in the context of fire.

Next examine the older main stems, which is the largest stem near the base of the plant (and above the burl if there is one). Look for either **smooth-reddish** or **rough-gray** bark. Older stems are most often smooth, exfoliating promptly after stem growth in the summer, to reveal older stems varying from dark purplish reds or brick reds to brighter orange red. In the foothills of the Sierra and along the California coast, a number of species exhibit bark that is usually grayish in color, that becomes rough with flat, shredded pieces of bark with age. This type of variation occurs among species in *Arbutus* (madrone) as well.

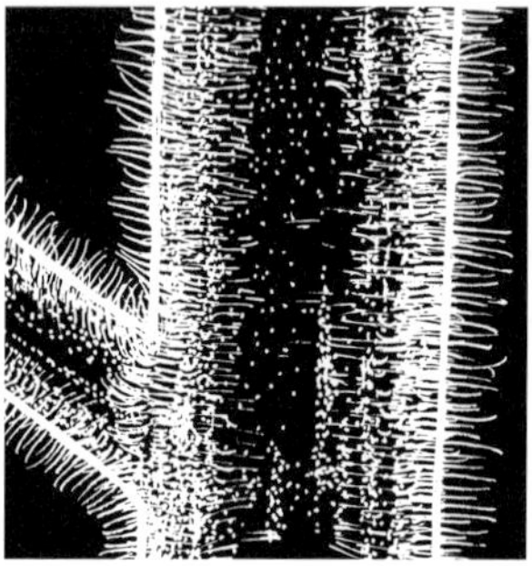

pubescent

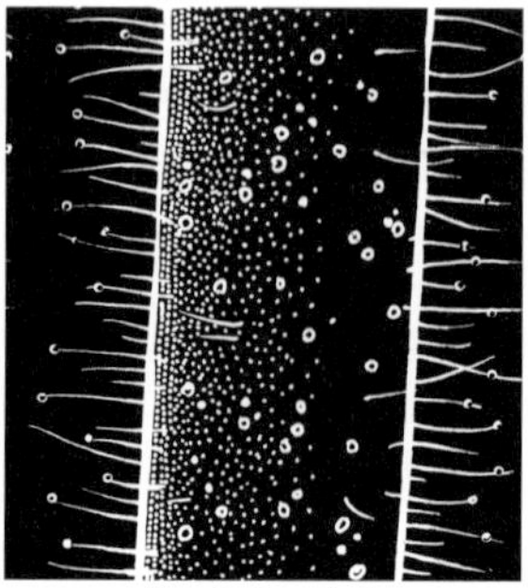

glandular-hairy

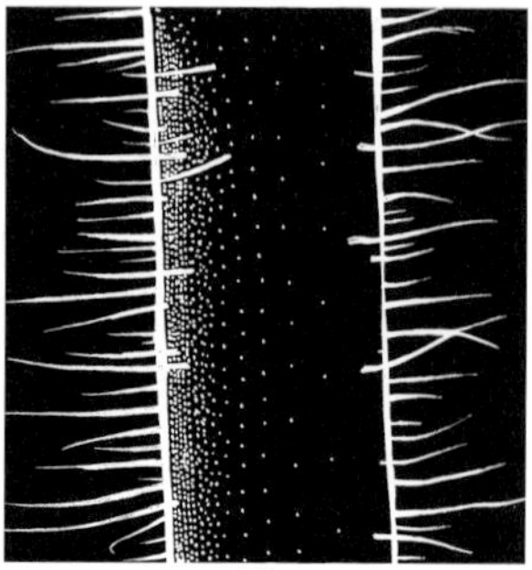

stiff hairs

In contrast, young stem-twigs can be quite different. Current year growth may possess an array of different characteristics involving hairs of different lengths, multiple layers of hairs, presence or absence of glandular tips, or hairs that are stiff, wavy or softly matted. These variations are important in identifying certain taxa. For example, one important variation is whether the hairs have a small shiny drop of liquid-looking structures at the tips. Those 'drops' are called **glands**, and the hairs would be called **glandular hairs**. In some species, the color is important (*A. patula* has short glandular hairs at the inflorescence and the glands are golden). **Tomentose** stems have short hairs, often enough to cover the surface, while **canescent** stems are whitish, tangled with hairs long and thick enough to give the plant a whitish hue. **Pubescence**, or hairiness, is also found on leaves, inflorescence, and fruit.

LEAVES. Can be found on young stems and also possess variation in pubescence along with several other characteristics. Two general patterns of leaf morphology are found within *Arctostaphylos*. In some species, leaves are **bifacial** (two sides of the leaf are different). The differences are most often in coloring or pubescence patterns. In this book, you will sometimes see references to the top-side or to the bottom-side of the leaf.

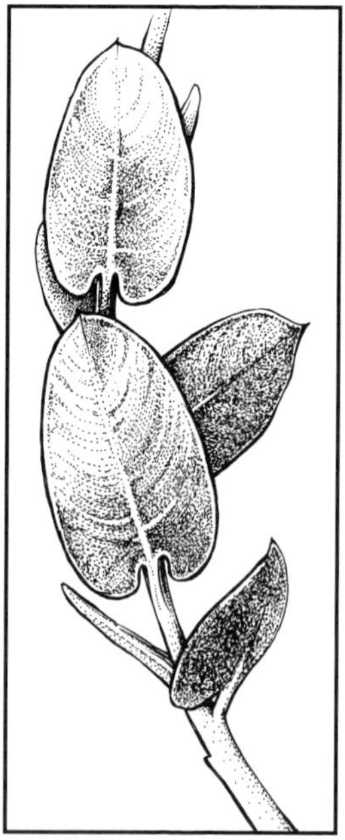

auriculate-clasping

For many species, learning how to identify stomata is important. **Stomata** refer to the small holes found in leaves that allow plants to bring in carbon dioxide for photosynthesis. Manzanitas have small stomata, but luckily they have a waxy build-up around the openings that make them easier to see as white dots using a 10x or 14x hand lens; stomata will appear as white spots crowding the areas between leaf veins. In bifacial leaves, stomata are almost always restricted to the bottom. More frequently, both the top and the bottom of the leaf have roughly equal numbers of stomata on each side and no variation in color. There are a few exceptions to these general patterns, which is why examining both leaf surfaces under magnification is important–although if the lower leaf is densely hairy obscuring stomata, focus on the upper and decide if it is the same or different than the lower (**isofacial** or bifacial). *Arctostaphylos insularis*, for example, has stomata restricted to the lower surface, while otherwise the leaf surfaces are similar in color and pubescence.

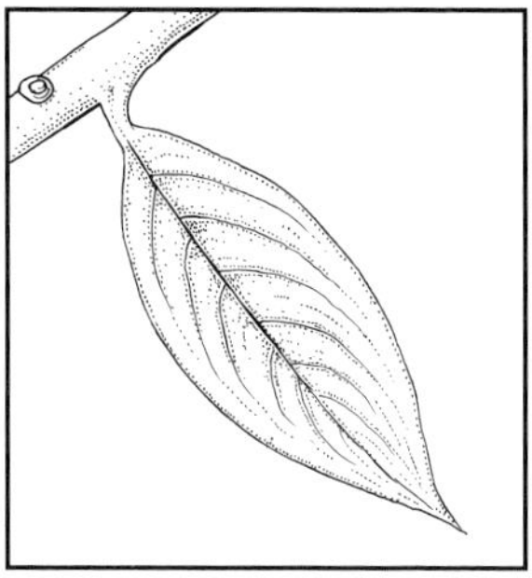

elliptic

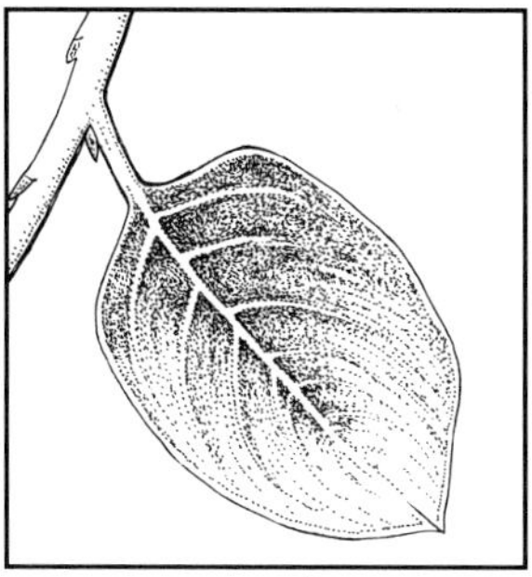

ovoid

Leaves also vary in shape. Broadly **elliptic** leaves (football shape) are common while one might also find **ovoid** (roundish) or **ovate** (larger at base, narrow at tip) leaves. A large group of taxa in the central California coast possess **auriculate** leaves, which are roughly ovate but with short **petioles** (leaf stalk). Next to the petiole they have a leaf that has expanded into 'earlobes' or auricles, such that the leaves appear to clasp the stems. These variations will not prove to be a difficulty, and each individual plant tends to have a general range of shapes, so the key only uses leaf shapes when the differences are large.

INFLORESCENCE. Manzanitas produce stem growth immediately following spring flowering. Immature inflorescences develop at the end of these new stems and generally remain dormant until the next flowering season, often 6 to 9 months later. Inflorescences are critical in identifying many taxa; either flowering or immature inflorescences are almost always present. Inflorescences range from short **racemes** to larger, spreading, many-branched **panicles**. Racemes refer to inflorescences with only a single main branch. In the keys, you'll sometimes see references to 'raceme-like' structures. What we mean by this is that most of the inflorescences are racemes, but frequently there can be a single small branch off the main branch, usually diminutive in comparison to the main branch. Usually, examination of a number of inflorescences will provide sufficient insight into whether the plant is raceme-like or has panicles (branched inflorescence). These structures should be examined under a hand lens for patterns of pubescence along the branches, bracts and pedicels of either the flower or fruit.

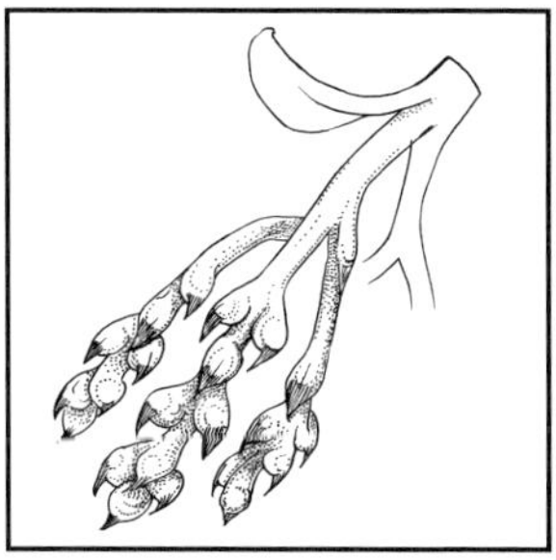

many-branched panicle

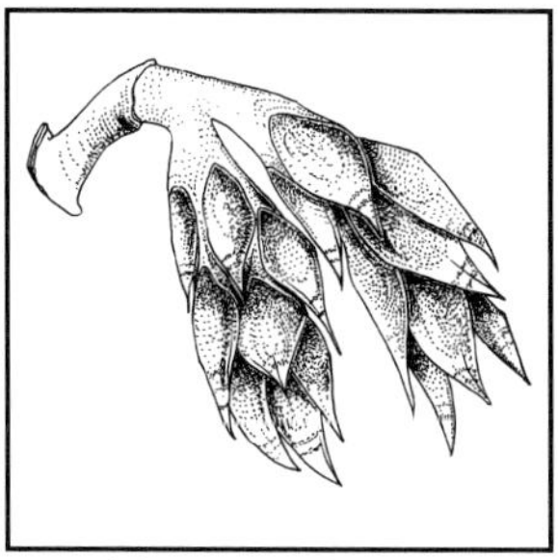

deltate to awl-shaped bracts

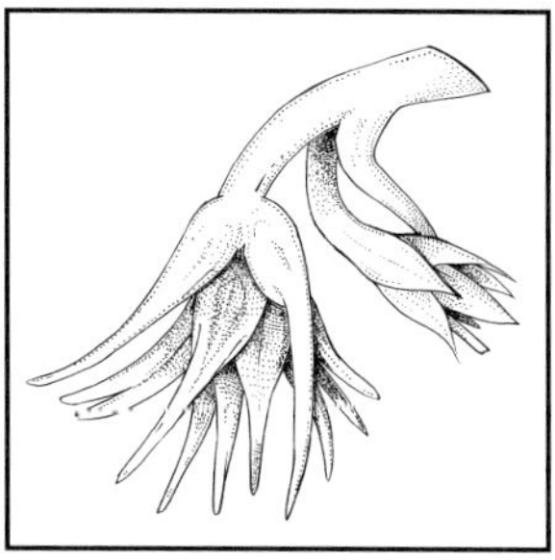

leafy-bract inflorescence

FLOWERS. While flower morphology does vary among taxa, it is rarely sufficient to aid in identification. One small lineage (*A. nummularia* & *A. sensitiva*) possesses 4-merous flowers (lobes, bracts, sepals, etc. in sets of 4) instead of the usual 5-merous, otherwise there is too much variation among individuals in shape and length for floral characteristics to be useful. One frequently important characteristic found in flowers is the pattern of pubescence associated with the ovary–if you remove the corolla (or fused petals) you will see the ovary. Frequently, the pubescence condition of the ovary is similar to that of mature fruit, but in many taxa this is not the case. For example, *A. canescens* has densely pubescent ovaries but the fruits often mature to a **glabrous** (bald or hairless) condition. There are a few taxa for which ovary pubescence is the key distinguishing character in keys and plant distribution can also be used in those cases.

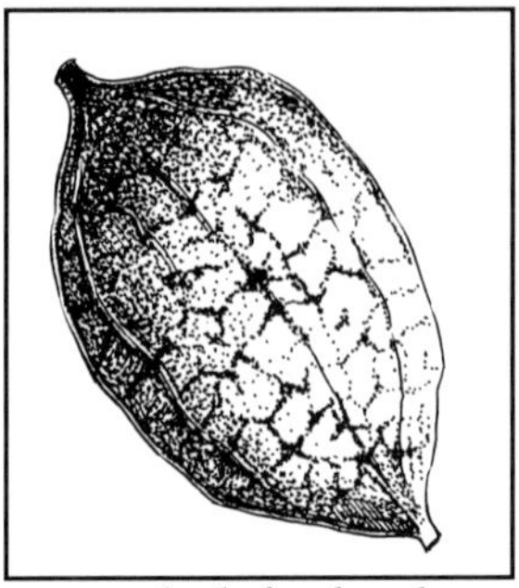

completely fused nutlet

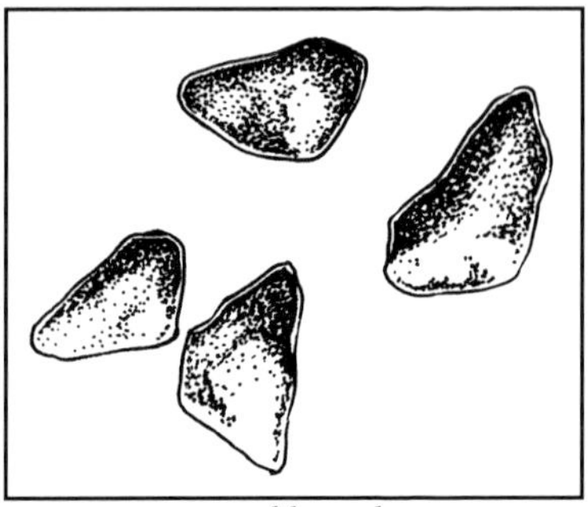

separable nutlets

FRUITS. A critical characteristic for identifying many of the taxa in this genus, fruits are produced early in the spring, mature in the summer and persist well into the fall and winter, and often remain on the plants into the next flowering season. Fruits are usually always around to be used in identification. They can also be found in the soil-litter during the latter part of the season.

The fruits are a type of **drupe**, which means the fruit covers a nut-like structure with the seed. The **exocarp** (skin of the fruit) is dry at maturity, the **mesocarp** (meat of the fruit) is dry and mealy or absent, and the fruit itself contains multiple hard **nutlets** or **stones**. Each stone is really an **endocarp** (hardened inner part of the fruit shaped like a segment of an orange) that surrounds a single seed. Nutlets are separate in many taxa. In some, however, nutlets may be fused (as many as 2 or 3) while other nutlets in the same fruit remain variously distinct (free). In contrast, some species characteristically possess fruit within which all nutlets are fused together yielding a single structure, usually quite spherical or elliptical in shape. With only a few exceptions, fruit fall into two general shapes: **spherical-globose** versus **depressed-globose**. Depressed-globose look like small apples because they are dimpled on both poles (hence the common name manzanita or "little apple"). Fruit also vary in pubescence patterns. They can be smooth and glabrous, pubescent, or viscid (sticky). Size, fruit shape, pubescence patterns, and condition of the nutlets are all critical characters for distinguishing among some groups of species.

Finally, what happens if you're out keying at a time of the year when one part of the key asks for characteristics that are not present on the plant? A good approach is to skip over some pairs of choices and go 'both ways' in the key. With these simplified keys, it should become clear what is the right direction. Also, look for distribution data in the key or species treatments.

Remember keying is more like an art than a black-and-white quantitative science. Use whatever information available to come to the most logical conclusion and, when you return from the field, look up photos and other resources on the Jepson Interchange. Like playing a musical instrument, practice yields proficiency and your skill at identifying manzanitas will eventually make sweet music indeed.

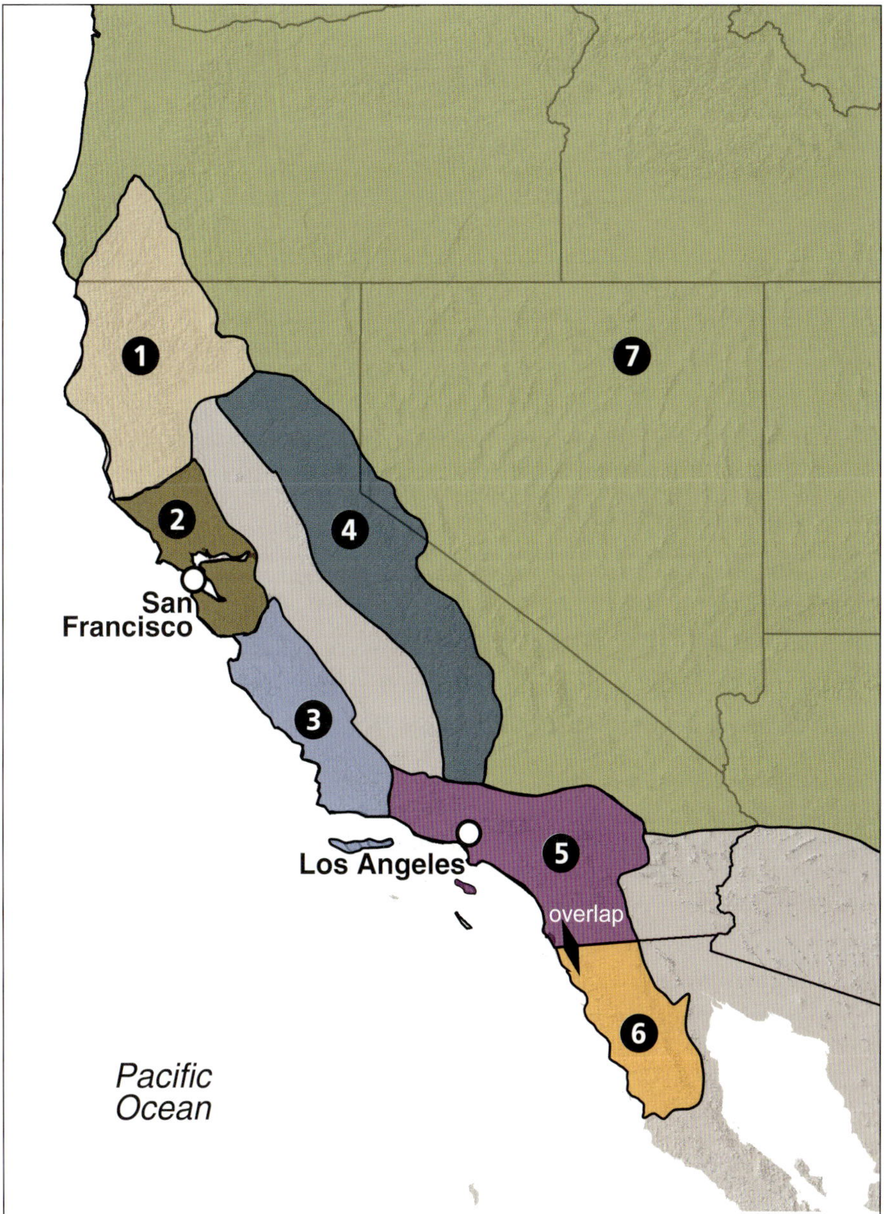

Keys By Region

1 Klamath Mountains (pg. 28)
2 San Francisco Bay (pg. 30)
3 Central Coast (pg. 33)
4 Sierra Nevada (pg. 37)
5 Southern California (pg. 38)
6 Baja California (pg. 39)
7 Outside the California Floristic Province (pg. 41)

A note before keying:

Keys all have a simple sequence to them but try using multiple keys to get better at using them. First, check the stems at the base of several individual shrubs to decide if there is a burl or not. While you're looking for burls, note whether the large, older stems are smooth reddish, blushed, or rough and grey-shreddy. Next, examine the terminal leafy, young branchlets for patterns of hair, whether they are present or absent, have short or long hairs (or both!), and whether any of the hairs are glandular or not. Next, look to the top and bottom of the leaves and if they are the same or different. If you have a hand lens, look for small, white dots on the two leaf surfaces. Examine the inflorescences (in either bud, flowering, or in fruit) for their pubescence as well, but also note whether branches of the inflorescence are simple racemes, i.e. no or one branched; or, if they are multi-branched, panicle-like inflorescence. Also, whether the inflorescence bracts are generally flat and leaf-like or small and scale-like. You'll get the pattern of these key sequences as you work through one or another key over and over (believe us ... this is never boring!). One way to help at the beginning is to mark the species you know are in the county in which you're standing. There are lists in the back. This will help at the beginning for you to see some of the distinctions and get used to the 'language' of the keys. Keys are art as much as science, and manzanitas are distinctly variable, so be comfortable with that variation.

Klamath Mountain Region Key:

1. Burls at base of main stem or where stems root, pl sprouting after fire
  2. Pl prostrate to mounded; infl raceme, 0–1-branched; fr dark brown; (gaps in high elevation conifer forest, nw coastal mnts) ***...A. nevadensis* subsp. *knightii*** (may extend north into Klamath Mountains in Oregon)
  2' Pl erect; infl with multiple branches (panicle)
    3. Twig and/or infl axes and/or bracts glandular hairy
      4. Glands black or clear (pink), on short and long hairs; burl prominent; lvs green to grey-green; (coast to interior chaparral) ...***A. glandulosa* subsp. *glandulosa***
      4' Glands golden, on short hairs; burl often flat, obscure; lvs bright green (montane chaparral) ... ***A. patula*[1] subsp. *patula***
    3' Twig, infl axes, bracts not glandular, either hairy or glabrous
      5. Twigs entirely glabrous; lvs glaucous white, fr 12-16 mm in diameter, dark chocolate brown (reported from e Siskyou Co. near Shasta) ... ***A. mewukka* subsp. *mewukka***
      5' Twigs hairy; lvs green; fr 8-12 mm in diameter, reddish brown
        6. Imm bracts scale-like or awl-shaped

1 Area covered by the
Key to the Klamath Mountain Region

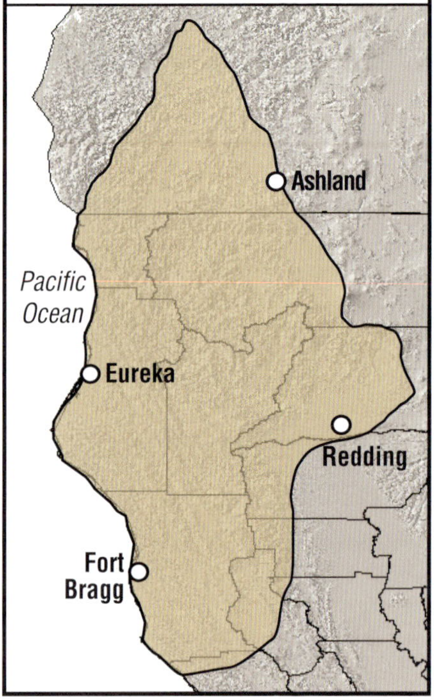

7. lvs bright green with light, short canescent hairs; burl often flat, obscure; frt not glandular (high elevation montane chaparral, conifer forest) ... ***A. patula*[2] subsp. *gankinii*** (probably confused with ***A. manzanita* subsp. *roofii*** in collection records)

7'. lvs dull green; burl prominent, fr may be glandular; (mid elevation chaparral, local, Glenn Co; possibly more widespread in inner coast range.) ..... ***A. manzanita* subsp. *roofii***

6'. Imm infl bracts at least occ lf-like (coast to interior, low to mid elevation, dense chaparral)... ***A. glandulosa* subsp. *cushingiana***

1' Burls 0 (or not visible); pl not sprouting after fire (if, in fact, no burl is present)

8. Lvs with stomata only on lower surface, surfaces gen different in color and/or hairiness

9. Fl parts in 4s; fr 3–4 mm wide, old st bark gen rough or shredding, grey or red-grey; c NCo (maritime chaparral, Mendocino Co. and n. Sonoma Co)... ***A. nummularia***

10. Pl 0.1–0.5 m; old st bark red-grey, gen rough or shredding; lf usually < 1.0 mm c NCo (Mendocino Co.) ..... **subsp. *mendocinoensis***

10' Pl 0.1–5 m; old st bark grey, shredding; lf > 1.2 mm (coastal prairie, maritime chaparral, closed cone forest, n Sonoma to c Mendocino) ..... ***subsp. nummularia*** (note: populations in coastal prairie at Stump Beach < .5m in ht, may be in other regional coastal prairies)

9' Fl parts in 5's; fr 5–15 mm wide; pl gen prostrate to low-mounded, lvs oblanceolate to obovate, occ narrowly elliptic, base wedge-shaped; ..... ***A. uva-ursi*** (low mounding individuals with lvs like ***A. columbiana*** with some stomata on upper surface are *A. x media*)

8' Lvs with stomata on both surfaces, surfaces similar in color and/or hairiness

11. Pl prostrate to prostrate-mounded, 0.1–0.5(3) m

12 Twig, infl axes, bracts glandular; lf glaucous grey (Scott Mtn Divide, Slate Mtn) ..... ***A. klamathensis***

12' Twig, infl axes, bracts gen nonglandular; lf green (widespread at high elevations on rocky outcrops and forest understory)..... ***A. nevadensis* subsp. *nevadensis***

11' Pl gen erect, occ mounded to erect, (0.1)0.3–8 m

13. Infl bracts fleshy, scoop-shaped or gen scale-like, deltate to awl-shaped, keeled or not

14. Lf ± white, white-glaucous, grey-green, grey-glaucous

15. Infl bracts glandular; infl "sticky" when pinched ..... ***A. viscida***

16. Ovary, fr rough, sticky-glandular; (common in serpentine chaparral) ..... **subsp. *pulchella***

16' Ovary, fr smooth, glabrous, not sticky; ..... **subsp. *viscid*a** (ne inner coast range in mid elevation chaparral and conifer forest)

15' Infl bracts glabrous, nonglandular or somewhat glandular-hairy (not sticky when "pinched");

17. Lf ± glaucous, glabrous or scabrous; ovary glabrous (or minutely glandular-hairy); (s outer coast range, Sonoma and Mendocino Cos) ..... ***A. manzanita* subsp. *glaucescens***

17' Lf blade canescent or densely white-tomentose, in age may become glabrous; ovary densely non-glandular hairy; (Red Mountain, Mendocino, Shasta, Trinity cos.) ..... ***A. malloryi***

14' Lf blade green, shiny or dull both surfaces, not ± white or grey due to ± glaucous or hairy surfaces

18. Twig glandular hairy

19. Imm infl spreading to erect, axes < 1 mm wide; bracts appressed; buds appearing as scattered beads infl branches

20. Lf blade 1.5–3 cm, dull green; twig finely glandular-hairy; (on serpentines in nw outer coast, local in Sonoma Co) ..... ***A. hispidula***

20' Lf blade 3–5 cm, grey green ; twig short-glandular-hairy; (inner coast range, s Mendocino, w Lake cos.) ..... ***A. stanfordiana* subsp. *raichei***

19' Imm infl pendent, axes gen > 1 mm wide; bracts gen ± appressed, > buds, buds not appearing as beads on infl branches

21. Glands of twig, imm infls gold; lf bright green, glabrous; (probably sprouting after fire, burl not visible; widespread high elevation conifer forest and montane chaparral) ..... ***A. patula***[3]

21' Glands of twig, imm infls dark or clear; lf dull green, scabrous, glandular;(upper mid elevation chaparral, Shasta, Tehama cos.) **..... *A. manzanita* subsp. *wieslanderi***

18' Twig nonglandular

22. Imm infl axis gen ≤ 1 mm wide; buds appearing as scattered beads; pl erect; (widespread in inner and outer coast range) **..... *A. stanfordiana* subsp. *stanfordiana***

22' Imm infl axis > 1 mm wide; buds not appearing as scattered beads on imm infl

23. Mound-like, spreading shrubs; twigs densely short-hairy (high elevation montane chaparral, conifer forest) ... ***A. patula* subsp. *gankinii***

23'. Mounding or often erect shrubs; twigs smooth to finely short-hairy (Low to moderate elevation chaparral, woodland ... ***A manzanita*** (in part)

24. Ovary, fr minutely glandular-hairy; pl erect; (chaparral, s inner coast range, often on volcanics) **..... *A. manzanita* subsp. *elegans***

24'. Ovary, fr glabrous; pl mounded or erect, often tree-like (widespread, foothill woodland, low elevation chaparral); **..... *A. manzanita* subsp. *manzanita*** (hybrids with ***A. stanfordiana*** resemble that species but have thicker imm infl)

13' Infl bracts gen lf-like, flat and prominent

25. Lf grey-canescent in youth, somewhat glabrous in age; corolla pink or white ..... ***A. canescens***

24. Ovary and pedicels nonglandular-hairy; (widespread, mid elevation chaparral) **..... subsp. *canescens***

24' Ovary and pedicels glandular-hairy; (less common but similar distribution as ***A. canescens***) ..... **subsp. *sonomensis*** (often occurs in same population as ***A. canescens***)

25'. Lf green or grey-green, glabrous to variously hairy but not grey-canescent

26.. Infl ± raceme, 0–1-branched, Pedicel, ovary, fr nonglandular-hairy; (500–940 m, nw KR, Del Norte Co.) ..... ***A. nortensis*** (may extend north into Oregon)

26'. Infl panicle, 3–10-branched, imm infl gen short- and long- nonglandular-hairy, occ glandular-hairy; (maritime chaparral and forest margins, common along coast to n. Sonoma Co) ..... ***A. columbiana***

San Francisco Bay Region Key:

1. Burls at base of main stem or where stems root; pl sprouting after fire

2. Lvs with stomata gen only on lower lf surface, surfaces gen differing in color and/or hairiness

3. Pl prostrate; burls at base of main st, sometimes where rooting; fr spheric; lf blades oblanceolate to obovate or narrowly elliptic; (San Bruno Mtn, Pt Reyes) ..... ***A. uva-ursi*** (includes *f. suborbiculata*, *f. marinensis*, and undescribed f. on San Bruno Mt)

3' Pl erect; burls at base of main st; fr depressed-spheric; lf blades oblong-ovate to lance-oblong;..... A. crustacea

4. Twig with short and long hairs, bottom of lvs with hair, may be lost with age

5. Lower side of lvs densely nonglandular-hairy (w Santa Cruz Mtns) ..... **subsp. *crinita***

5' Lower side of lvs ± nonglandular-hairy, in age glabrous (widespread) ..... **subsp. *crustacea***

4' Twig with short hairs, lacking long hairs, bottom of lvs glabrous (San Francisco, Lake Merced)..... **subsp. *rosei***

2' Lvs with stomata on both surfaces, occ fewer on the top, surfaces gen the same in color and/or hairiness (exc. *Arctostaphylos pacifica*)

6. Pl prostrate to mounded; infl raceme, 0–1-branched, lvs often serrate; fr ± red; lf surfaces differ in hue (local endemic on San Bruno Mtn) ..... ***A. pacifica***

6' Pl erect; infl panicle, lvs on mature pl generally smooth on edges (widespread)... ..***A. glandulosa***

7. Twig and/or infl axes and/or bracts glandular (widespread in n Bay and Mt Diablo) ..... **subsp. *glandulosa***

7' Twig, infl axes, bracts not glandular; (widespread in n Bay from coast to inner range)..... **subsp. *cushingiana*** (often mixed w/ subsp. *glandulosa*)

1' Burls 0; pl not sprouting after fire

8. Lvs with stomata on lower surface, surfaces gen different in color and/or hairiness

9. Flwr parts in 4s; fr 3–4 mm wide; old st bark smooth, ± red; w SnFrB (Bolinas Ridge, Mount Tamalpais, Santa Cruz Mtns) ..... ***A. sensitiva*** (fr "shatters" soon after maturity, fr hard to find on pl and litter)

9' Flwr parts in 5s; fr 5–15 mm wide

10. Pl gen prostrate to low-mounded, lvs elliptic to obovate (San Bruno Mtn, immediate coast)..... ***A. uva-ursi***

10' Pl erect, lvs oblong to triangular-ovate, auriculate or lobed at base

11. Old st bark persistent, shredding, gray or ± red; lf blue-green; 2-4 cm (low elevation maritime chaparral, Pajaro Hills) ..... ***A. pajaroensis*** (local endemic)

11' Old st bark smooth, ± red; lf light green; 4-7 cm (mid elevation forest margins, Santa Cruz Mtns) ..... ***A. andersonii***

8' Lvs with stomata on both surfaces, surfaces similar in color and/or hairiness

12. lvs oblong to triangular-ovate, auriculate or lobed at base

13. Pl prostrate to prostrate-mounded, 0.1–0.5(3) m; twig, imm infl, pedicel, ovary glandular; 200–400 m, (rocky maritime chaparral, San Bruno Mtn) ..... ***A. imbricata*** (local endemic)

13' Pl erect

14. lvs typically densely grey-canescent, may lose hair with age and seem grey-glaucous

15. Bracts on mature or imm inflorescence scale-like, somewhat fleshy, ft spheric to 15 mm diam, seeds fused (local endemic in Coulter pine chaparral, northern & central San Benito Mnts)... ***A. gabilanensis*** (local endemic)

15' Bracts on mature or imm inflorescence, flat and lf-like, ft depressed spheric, 8-12 mm diam, seeds separate

16. Twigs, ovary, ft non-glandular (Mount Diablo and n Contra Costa Co.) ..... ***A. auriculata*** (local endemic)

16' Twigs, ovary, ft glandular-hairy (n Ben Lomond Mtn, nw Santa Cruz Co.) ..... ***A. glutinosa*** (local endemic)

14'. lvs typically lacking dense hairs (glabrous), light green, grey green or glaucous-green

17. lvs generally ovate, 2.5-4.5 cm, generally roughly flat (maritime chaparral)
18. Lf blade bright green, not glaucous; twig long-glandular-hairy; (chaparral, widespread on Montara Mtn, few on San Bruno Mnt) ..... ***A. montaraensis*** (local endemic)
18' Lf blade glaucous-green; twig short-hairy w occasional long glandularhairs; (Sobrante, Huckleberry ridges; Alameda, Contra Costa cos.)..... ***A. pallida*** (local endemic)
17' lvs oblong-ovate, grey-green, 3-6 cm, boat-shaped; (maritime chaparral and forest edges, King's Mnt, and north, San Mateo Co.)..... ***A. regismontana*** (local endemic)
12'. lvs elliptic, ovate, oblong, but not clasping stem at base with lobes
19. Pl prostrate to prostrate-mounded, 0.1–0.5(3) m
20. Infl panicle, 3–5-branched; lf blade gen 3–5 cm (volcanic formations, ne Sonoma, nw Napa Cos.) ..... ***A. stanfordiana*** **subsp.** ***decumbens*** (local endemic)
20' Infl ± raceme or short panicle, 0–3(5)-branched; lf blade gen 1–3 cm
21. Fr 4–5 mm wide, pl essentially prostrate, lvs round to round-elliptic, (serpentine, San Francisco Presidio) ..... ***A. montana*** **subsp.** ***ravenii*** (local endemic)
21' Fr 6–8 mm wide, pl mounding, lvs round elliptic-elliptic (serpentine, n San Francisco) ..... ***A. franciscana*** (local endemic, nearly extinct)
19' Pl gen erect, 1-5 m or more
22. lvs grey-whitish green or glaucous-grey green, lvs glabrous, glaucous or canescent
23. Bracts of imm or mature infl bract-like, although they may seem fleshy in some species, awl-shaped or apiculate, keeled or not
24. Lf ± white, white-glaucous, gray-green, gray-glaucous
25. Fr (10)12–16 mm wide, spheric or ± spheric;, nutlets fused together; lf glabrous;(mostly serpentine in Bay Area, widespread)..... ***A. glauca***
25' Fr (6)8-10 mm wide, depressed spheric, nutlets free or mostly free
26. Imm infl bracts glandular-viscid; sticky when pinched ..... ***A. viscida***
27. Ovary, fr rough, sticky-glandular; (common on serpentines in n Bay) ..... **subsp.** *pulchella*
27' Ovary, fr smooth, glabrous, not sticky; (rare, non-serpentine, Marin) ..... **subsp.** ***viscida***
26' Imm infl bracts glabrous, sparsely glandular-hairy, not sticky when pinched; (Sonoma Co.) ..... ***A. manzanita*** **subsp.** ***glaucescens***
23' Bracts of imm or mature infl flat and lf-like, regardless of color
28. Ovary glabrous; corolla white; infl 0–1-branched, ± raceme; pedicel 5–7 mm; (s Santa Cruz Mtns) ..... ***A. silvicola*** (local endemic)
28' Ovary glandular or nonglandular-hairy; corolla pink or white ..... ***A. canescens***
29. Ovary nonglandular-hairy; (common, mid elevation chaparral, s Sonoma and Napa, Mt Tamalpais in Marin, s Santa Cruz Co) ..... ***subsp. canescens***
29' Ovary glandular-hairy; (occasional, distribution similar as *A. canescens*) ..... ***subsp. sonomensis*** (occurs within populations of *A. canescens*)
22' lvs bright green, green or light green, not grey-whitish or densely canescent
30. Imm infl bracts gen scale-like, deltate to awl-shaped, keeled or not
31. Infl gen ± raceme, occ 1–2-branched, imm often spheric, cup-like, bracts ± spreading; (s Santa Cruz Co) ..... ***A. hookeri*** **subsp.** ***hookeri*** (local endemic)
31' Infl panicle, (1)3–8-branched
32. Twig glandular-hairy, either densely glandular-hairy, or with occasional short glandular hairs, sometimes below longer nonglandular hair
33. Imm infl spreading to erect, axes thin ($\leq$ 1 mm wide); bracts appressed and short, ± = buds; buds appearing as scattered beads (Rockpile Road, s Sonoma Co.) ..... ***A. hispidula***

33’ Imm infl pendent, axes gen wider (> 1 mm wide); bracts gen ± appressed, > buds, buds not appearing as bead

34. Twig with some long glandular hairs; siliceous shale, sw SnFrB (n Ben Lomond Mtn, nw Santa Cruz Co.) ..... ***A. ohloneana*** (local endemic)

34’ Twig with dense to sparse glandular hairs or sessile glands; serpentine, (Sonoma Co.) ..... A. bakeri

35. Twig with glandular short hairs; (central Sonoma Co. s of Guerneville) ..... **subsp. *bakeri*** (local endemic)

35’ Twig with sessile glands; (Sonoma Co. from The Cedars to Healdsburg) ..... **subsp. *sublaevis*** (local endemic)

32’ Twig nonglandular

36. Imm infl axis gen narrow (≤ 1 mm wide)

37. Lf blade 1–2.5 cm; bracts > buds; buds not appearing as beads; pl mounded to erect; (Vine Hill Rd. near Forestville, Sonoma Co.) ..... ***A. densiflora*** (a local endemic)

37’ Lf blades 3–5 cm; bracts ± = buds; buds appearing as scattered beads; pl erect; (widespread, Napa, Sonoma, Lake Cos.) ..... ***A. stanfordiana* subsp. *stanfordiana***

36’ Imm infl axis wider > 1 mm wide

38. Lf blade 1–2.5 cm, pl short, gen up to 1.5 m, infl axis densely short hairy ..... ***A. montana* subsp. *montana*** (local endemic)

38’ Lf blade 1.5–5 cm, pl gen tall, up to 8 m; infl axis lightly hairy to glabrous

39. Ovary, fr minutely glandular-hairy; pl erect;..... ***A. manzanita* subsp. *elegans***

39’ Ovary, fr glabrous; pl mounded or erect;

40. Lf blade bright green; infl 2–4-branched, axis > 1 mm wide; imm infl branches to 15 mm; (Mount Diablo) .....***A. manzanita* subsp. *laevigata*** (local endemic)

40’ Lf blade dull to bright green; infl 4–7-branched, axis ± 1 mm wide; imm infl branches gen 15–25 mm; (widespread in n Bay, frequently in oak woodlands, to e Santa Clara, not in Santa Cruz Mnts)..... ***A. manzanita* subsp. *manzanita***

30’ Imm infl bracts gen lf-like, flat

41. Infl ± raceme, 0–1-branched, twig, infl axes, bracts glandular; imm infl obscure (Marin Co.) ..... ***A. virgata*** (local endemic)

41’ Infl panicle, 3–10-branched, lf lanceolate to ovate, dark green; imm infl obvious, gen short- and long- nonglandular-hairy, occ glandular-hairy (coastal n Sonoma, widespread along coast to coast)..... ***A. columbiana***

Central Coast Manzanita Key

1. Burls at base of main stem, also above or not; pl sprouting after fire

2. Lvs with stomata gen only on lower, surfaces gen differing in color and/or hairiness

3. Old stem bark persistent, gray, shredding ..... ***A. tomentosa***

4. Twig with glands (Jack’s Peak and vicinity, Monterey Co) ..... **subsp. *bracteosa*** (local endemic)

4’ Twig without glands

5. Twig short-tomentose and long-nonglandular-hairy (near Morro Bay, San Luis Obispo Co.) ..... **subsp. *daciticola*** (local endemic)

5’ Twig short-tomentose or short-nonglandular-hairy

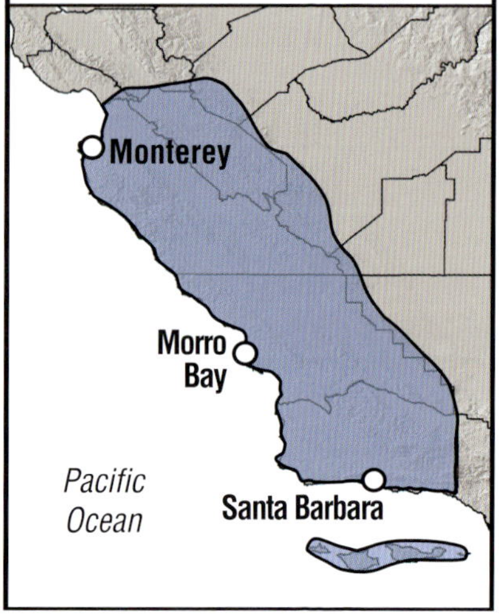

3 Area covered by the **Key to the Central Coast Region**

6. Lvs tomentose abaxially (Monterey Co. to n San Luis Obispo Co.) ..... **subsp. *tomentosa***

6' Lvs glabrous abaxially (Jack's Peak and vicinity, Monterey Co.)..... **subsp. *hebeclada*** (local endemic)

3' Old stem bark gen smooth or peeling, ± red..... ***A. crustacea***

7 Twig, nascent infl axes densely glandular-hairy (Santa Cruz and Santa Rosa Islands, Santa Barbara Co) ..... **subsp. *subcordata*** (occasional individuals of *A. crustacea* elsewhere in its range)

7' Twig, nascent infl axes glabrous or nonglandular-hairy, occ sparsely glandular-hairy

8. Twig with short and long hairs

9. Lower lf surface densely non-glandular-hairy (sw San Mateo Co. and w Santa Cruz Co., Santa Barbara) ..... **subsp. *crinita***

9' Lower lf surface ± moderately hairy to non-hairy (widespread along Coast Ranges from Napa Co. to Channel Islands) ..... **subsp. *crustacea***

8' Twig with short hairs only

10. Lower lf surface densely short non-glandular hairy (Santa Cruz and Santa Rosa Islands) ..... **subsp. *insulicola*** (local endemic)

10' Lower lf surface glabrous

11. Pedicel, ovary glabrous (hills near Burton Mesa, Santa Barbara Co.) ..... **subsp. *eastwoodiana*** (local endemic)

11' Pedicel, ovary gen short non-glandular hairy (along edge of Big Sur Coast, Monterey Co., and San Francisco) ..... **subsp. *rosei***

2' Lvs with stomata on both surfaces, occ fewer above, surfaces gen the same in color and/or hairiness

12. Old stem bark persistent, shredding, gray (maritime chaparral, Nipomo, Burton mesas, Point Sal; sw San Luis Obispo, nw Santa Barbara cos.) ..... ***A. rudis***[1] (local endemic)

12' Old stem bark smooth or peeling, ± red ..... ***A. glandulosa***

13. Twigs or inflorescence with glandular hairs

14. Twig gen glandular; (widespread) ..... **subsp. *glandulosa***

14' Twig not glandular, infl glandular; (Santa Lucia Range) ..... **subsp. *howellii***

13' Twigs and inflorescence non-glandular

15. Twig hairs bristly with prominent long hairs (western Transverse Range, Santa Barbara Co.) ..... **subsp. *mollis***

15' Twig hairs sparsely short hairy to canescent with dense short hairs

16. Fr with separable nutlets, some fusion, stems with gray, dense short hairs (widespread in coastal mountains) ..... **subsp. *cushingiana*** (sometimes occurs in same population as subsp. *glandulosa*)

16' Fr with nutlets fused (rarely breaking apart to 2 or 3 pieces) (San Rafael Mtns; Santa Barbara Co.) ...**subsp. *gabrielensis***

1' Burls 0; pl not sprouting after fire

17. Lvs with stomata only on lower lf surface, lf surfaces gen different in color and/or hairiness

18. Lower leaf surface gen tomentose, sparsely so in age

19. Plants prostrate to mounded, lf blade narrowly obovate to oblanceolate, 1–2 cm, base wedge-shaped; (maritime chaparral, sandy habitats around Monterey Bay, Monterey Co.) ..... ***A. pumila*** (local endemic)
19' Plants erect, lf blade oblong-ovate to -elliptic, 1.5–3 cm, base ± truncate to cordate; (maritime chaparral, s Morro Bay, San Luis Obispo Co.) ..... ***A. morroensis*** (local endemic)
18' Lower leaf surface glabrous or sparsely non-glandular hairy
20. Pl gen prostrate to low-mounded
21. Lvs oblanceolate to obovate, occ narrowly elliptic, base wedge-shaped;(Pfeiffer Beach, nw Monterey Co., on sea stack) ..... ***A. uva-ursi***
21' Lvs elliptic to round-ovate, base truncate to ± lobed; (maritime chaparral near the ocean, nw Big Sur coast, Monterey Co.) ..... ***A. edmundsii*** (local endemic)
20' Pl erect
22. Petioles 4–8 mm; lf oblong-elliptic, base rounded, tip obtuse; older stems smooth, reddish, with glaucous patches (maritime chaparral, Santa Cruz Island) ..... ***A. insularis*** (local endemic)
22' Petioles 1–2(4) mm; lf oblong to triangular-ovate, base lobed, tip acute; older stems with grey, shreddy bark (maritime chaparral, n Monterey Co. Pajaro Hills) ..... ***A. pajaroensis*** (local endemic)
17' Lvs with stomata on both surfaces, surfaces similar in color and/or hairiness
23. Pl prostrate to prostrate-mounded, 0.1–0.5(3) m
24. lf base lobed, clasping; imm infl bracts lf-like (maritime chaparral, s Monterey, nw San Luis Obispo cos.) ..... ***A. cruzensis*** (local endemic)
24' lf base narrowly wedge-shaped, not clasping; nascent infl bracts scale-like (maritime chaparral, San Luis Obispo Co.) ..... ***A. hookeri* subsp. *hearstiorum*** (local endemic)
23' Pl gen erect, occ mounded to erect, (0.1)0.3–8 m
25. Nascent infl bracts gen scale-like or fleshy, deltate to awl-shaped, keeled or not, fleshy or not
26. Leaves green, infl gen ± raceme, occ 1–2-branched, fruit <10-12 mm wide, stones gen separate
27. Old stem bark persistent, shreddy, gray — (maritime chaparral, Nipomo, Burton mesas, Point Sal; sw San Luis Obispo, nw Santa Barbara cos.) ..... ***A. rudis***[2] (local endemic)
27' Old stem bark smooth, ± red
28. Infl a panicle, nutlets fused together in the fr
29. Leaves elliptic (upper elev mnts eastern Santa Barbara Co)...***A. parryana* subsp. *parryana***
29' Leaves lobed, auriculate, nearly clasping stem (Santa Ynez Mts) ...***A. refugioensis***
28' Infl +/- a raceme, nutlets separable in fr
30. Imm infl club-like, bracts recurved; (chaparral, widespread species in general; in central coast, Los Padres and San Benito Mountains) ..... ***A. pungens***
30' Imm infl spheric, cup-like, bracts ± spreading; (maritime chaparral, hills surrounding Monterey Bay, s Santa Cruz and n Monterey Cos.) ..... ***A. hookeri*** subsp. hookeri (local endemic)
26' Leaves ± white hairy or white-glaucous, infl panicle, (1)3–8-branched, fruit large (10) 12-16 mm wide, ± spheric, stones fused
31. Petiole 7–15 mm; lf glabrous, base rounded, truncate, or ± lobed; ovary, fr glandular-sticky; (chaparral, conifer woodland, widespread interior mountains) ..... ***A. glauca***
31' Petiole 1–4 mm; lf gen short non-glandular hairy, base distinctly lobed; ovary, fr glabrous, not sticky (n&c Gabilan Range, San Benito, Monterey cos.) ..... ***A. gabilanensis*** (local endemic)

25' Nascent infl bracts gen lf-like, flat

32. Old stem bark persistent, gray, shredding (exc occ smooth), leaves dark green, base lobed, clasping; s CCo (maritime chaparral, volcanic hills, w Los Osos Valley, San Luis Obispo Co.) ..... ***A. osoensis*** (local endemic)

32' Old stem bark smooth, ± red, leaves various

33. Lf gray-canescent in youth, glabrous in age

34. Lf base lobed, clasping; petiole to 4 mm (chaparral, se of Cuesta Pass, San Luis Obispo Co.) ..... ***A. luciana*** (local endemic)

34' Lf base wedge-shaped, rounded, truncate, or ± lobed, not clasping; petiole 3–10 mm

35. Ovary glabrous; corolla white (chaparral, gen on serpentine, c&s Santa Lucia Range, n San Luis Obispo and s Monterey Cos.) ..... ***A. obispoensis***

35' Ovary glandular or non-glandular hairy; corolla pink or white (unconfirmed in n Monterey Co., near by in Santa Cruz Co.) ..... ***A. canescens***

36. Ovary non-glandular hairy; (widespread, common in montane chaparral)..... **subsp. *canescens***

36' Ovary glandular-hairy; (similar to *A. canescens*) ..... **subsp. *sonomensis*** (often occurs within populations of *A canescens*)

33' Lf green to glaucous, glabrous to variously hairy but not gray-canescent

37. Lf base lobed (truncate to lobed in *Arctostaphylos viridissima*), often clasping (exc *Arctostaphylos hooveri* with longer petiole)

38. Imm infl, twigs non-glandular hairy

39. Pedicel non-glandular hairy; n ChI (maritime chaparral, e Santa Cruz Island) ..... ***A. viridissima*** (local endemic)

39' Pedicel glabrous or sparsely non-glandular hairy;

40. Lf blade 2–5 cm; fr 8–12 mm wide; stones gen fused; (maritime chaparral, Morro Bay to Avila Beach, San Luis Obispo Co.) ..... ***A. pechoensis*** (local endemic)

40' Lf blade 1–2.5 cm; fr 5–8 mm wide; stones free; (maritime chaparral, w Santa Barbara Co.) ..... ***A. purissima* subsp. *purissima*** (local endemic)

38' Imm infl, twigs glandular-hairy

41. Fr 10–15 mm wide, spheric; stones fused; petioles short, lf base clasping stems or nearly so (chaparral, Santa Ynez Mtns)..... ***A. refugioensis*** (local endemic)

41' Fr 6–10(15) mm wide,

42. Fr depressed-spheric; stones free; petioles 3–6 mm; lvs partially lobed but not clasping; (chaparral, high coastal ridges, Santa Lucia Range) ..... ***A. hooveri***

42' Fr spheric; stones frequently fused together, petioles <2mm; lvs lobed, strongly clasping (gen south of Lompoc and Hwy 246, west of Hwy 101; Santa Ynez Mts. ...***A. purissima* subsp. *globosa***

37' Lf base rounded, truncate, wedge-shaped, or ± lobed, not clasping

43. Infl ± raceme, 0–1-branched, lvs spaced, pl erect (Pismo Beach vicinity, s Santa Lucia and La Panza ranges) ..... ***A. pilosula***

43' Infl panicle, 3–10-branched, twigs, infl, glandular-hairy; pl prostrate to mounded or erect

44. Lvs overlapping; pl prostrate to mounded or sometimes erect, 0.1–2 m; (Santa Rosa Island, Santa Barbara Co.) ..... ***A. confertiflora*** (local endemic)

44' Lvs spaced along twigs, stegs, infl glandular hairy, plants 1-3 m; (area around Monterey Bay, Ft. Ord to Toro Mt.) ...***A. montereyensis*** (local endemic)

Sierra Nevada Region Key:

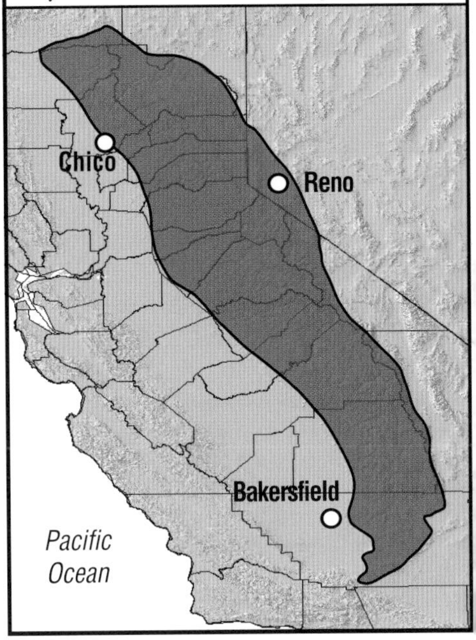

1. Pl with burl at the st base, resprouting after fire
 2. Lvs glaucous white; fr large, dark mahogany-brown, twigs glabrous (mid-elev, mostly central Sierra) ..... ***A. mewukka* subsp. *mewukka***
 2' Lvs green; fr smaller, reddish to brown, twigs with hairs, glandular or not
  3. Lvs bright green or slightly grey-green; burl often flat, obscure; ..... ***A. patula***
   4. Twig hairs usually with minute golden glands (montane chaparral, widespread at high elev) ..... ***A. patula*[1] subsp. *patula***
   4' Twig hairs lacking minute golden glands, instead with dense short white hair (montane chaparral, Sierra at high elev) ..... ***A. patula*[2] subsp. *gankinii*** (generally occurs within populations of glandular *A. patula*)
  3. Lvs dull to bright green; burl prominent; (chaparral, mid elevation, foothill woodland zone, Butte Co.) ..... ***A. manzanita* subsp. *roofii*** (more common across Sacramento Valley in inner coast range; subject to confusion with non-glandular *A. patula*)

1' Pl lacking a burl at st base; not resprouting after fire
 4. Lvs with stomata only on lower lf surface, surfaces gen different in color and/or hairiness, pl gen prostrate to low-mounded; lvs oblanceolate to obovate;(above Convict Lake, Mono Co) ..... ***A. uva-ursi*** (f. *adenotricha*, only location in California)
 4' Lvs with stomata on both surfaces, surfaces similar in color and/or hairiness
  5. Pl prostrate to prostrate-mounded, 0.1–0.5(3) m, twig sparsely short non-glandular hairy; bracts near infl base occ 5–10 mm; (widespread at high elev on outcrops or conifer forest understory) ..... ***A. nevadensis* subsp. *nevadensis***
  5' Pl gen erect, occ mounded to erect, (0.1)0.3–8 m
   6. Leaves glaucous–white or canescent whitish-green
    7. Old stem bark grey, with persistent shreds of old bark; leaves typically canescent, bracts lf-like but white canescent (shale barren chaparral, mid-elevation, Placer, El Dorado, Tuolumne Cos.) ..... ***A. nissenana*** (local endemic)
    7' Old stem bark smooth red, leaves smooth waxy-glaucous, sometimes glandular, bracts gen scale-like, deltate to awl-shaped, keeled or not
     8. Fr (10)12–16 mm wide, spheric or ± spheric, stems, nascents, fruit glabrous and non-glandular; nascent infl bracts appressed; fr glabrous;(mid elevation chaparral near Paradise, Butte Co.) ..... ***A. mewukka* subsp. *truei*** (local endemic)
     8' Fr 6–12 mm wide, depressed-spheric, stems, nascents or fruits glandular-viscid or glandular hairy; imm infl bracts glandular; (widespread) ..... ***A. viscida***
      9. Twig densely glandular-hairy; lf sparsely short-glandular-hairy, ciliate, papillate, scabrous; ovary, fruit viscid. .....***A. viscida* subsp. *mariposa***
      9' Twig glabrous to sparsely short-nonglandular-hairy; lf glabrous; ovary, fr smooth, glabrous, not sticky; ..... ***A. viscida* subsp. *viscida***

6' Leaves green, shiny or dull

10. Infl gen ± raceme, occ 1–2-branched; old st bark smooth, dark red with ± gray or ± glaucous patches; twig glandular-hairy; plants 0.5-1.5 m; (chaparral, Amador, Calaveras Cos. on oxisols or acidic soils) ..... ***A. myrtifolia*** (local endemic)

10' Infl a panicle, (1)3-8 branched; plants tall, 2-8 m, old st bark smooth reddish

11. Twigs glandular

12. Glands of twig, imm infls golden on short hairs; lf bright green, glabrous; (widespread at higher elev)..... ***A. patula*[3] subsp. *patula*** (almost all *A. patula* in the Sierra can resprout after fire but burls are obscure and can be missed)

12' Glands of twig, imm infls dark or clear dense, long hairs; lf dull green, scabrous, glandular; (Shasta, Tehama cos.) ..... ***A. manzanita* subsp. *wieslanderi***

11' Twig nonglandular; lf blade dull to bright green

13. Erect, often often tree-like (chaparral or foothill woodland, widespread at low elev) ..... ***A. manzanita* subsp. *manzanita***

13'. Spreading, mound-like shrub (montane chaparral, conifer forest at high elevations)..... ***A patula*[3] subsp. *gankinii***

Southern California Key:

1. Burls at base of main st; pl sprouting after fire

2. Twig and/or infl axes and/or bracts glandular

3. Glands black or clear (pink), on short and long hairs; burl hemispheric; ..... ***A. glandulosa*** (in part)

4. Lvs green to gray-green ..... **subsp. *glandulosa***

4' Lvs strongly white-glaucous ..... **subsp. *leucophylla***

3' Glands golden, on short hairs; burl often flat, obscure; ..... ***A. patula*[1]**

2' Twig, infl axes, bracts not glandular (exc twig minutely glandular, infl axes sparsely short glandular in Arctostaphylos rainbowensis)

5 Area covered by the **Key to the Southern California Region**

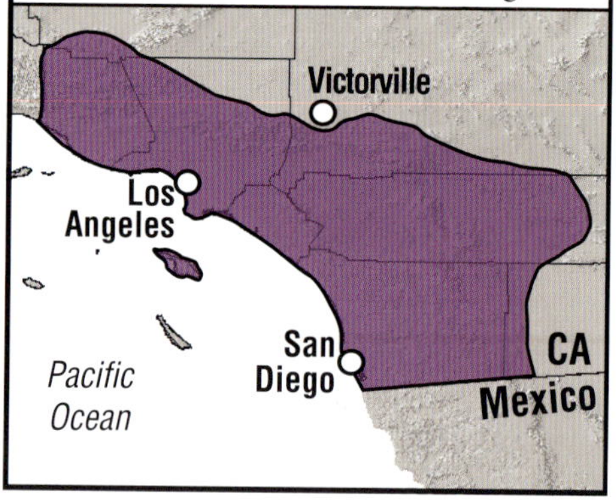

5. Fr spheric

6. Twig with short hairs, minutely glandular; (San Diego, Riverside cos.) ..... ***A. rainbowensis***

6' Twig with short non-glandular hairs; SnBr, (chaparral and pine woodland, San Jacinto, Santa Rosa, San Ysidro mtns) ..... ***A. parryana***

7. Lvs gray-glaucous; (San Jacinto, Santa Rosa, San Ysidro mtns) .... ***A. parryana* subsp. *desertica***

7' Lvs green; ..... ***A. parryana* subsp. *tumescens***

5' Fr depressed-spheric ..... ***A. glandulosa*** (in part)

8. Lvs strongly white-glaucous; (chaparral, SE and SW interior and coastal mnts) ..... **subsp. *adamsii***

8' Lvs green, yellow-, or gray-green;

10. Twig hairs soft, dense, and long (Transverse Range, n Los Angeles and Riverside Cos.)..... **subsp. *mollis***

10' Twig hairs not soft, dense, and long

11. Fr depressed-spheric, stones fused [or rarely in 2(3) fused units] (west Transverse Range, San Gabriel Mountains) ..... **subsp. *gabrielensis***

11' Fr depressed-spheric, stones gen free

12. Lf blades green, margins often ± red; fr markedly depressed-spheric; (coastal San Diego Co.) ..... **subsp. *crassifolia*** (local endemic)

12' Lf blades green to gray-green; fr not markedly depressed-spheric; (widespread coastal mountains) ..... **subsp. *cushingiana***

1' Burls lacking from main st base; pl not sprouting after fire

13. Fr spheric, stones fused into a solid unit

14. Lf distinctively white to gray-green, fr large (10)12–16 mm wide, fr sticky; (chaparral, woodland; often tree-sized; widespread in region at mid elevations) ..... ***A. glauca***

14' Lf green to glaucous-green, fr smaller (6-12 mm), fr not sticky

15. Twigs densely glandular-hairy, lf-like pink bracts soon deciduous, glaucous-green leaves (tall shrub, chaparral, eastern Peninsular Range s to Baja) ..... ***A. pringlei* subsp. *drupacea***

15' Twigs with short non-glandular hairs, bracts scale-like, persistent, leaves shiny green (mounding shrub, open pine woodlands, high elevations in the Transverse Range) ..... ***A. parryana* subsp. *parryana***

13' Fr depressed-spheric, stones separable

16. Imm infl bracts scale-like, lv bright green

17. Infl gen ± raceme, occ 1–2-branched, imm infl club-like, bracts recurved; (chaparral, widespread, mid elevation) ..... ***A. pungens***

17' Infl a spreading panicle, 3-8 branched, bracts appressed to buds (montane chaparral, widespread, high elevation) ..... ***A patula***[2] (typically resprouting but burl may be obscure)

16' Imm infl bracts lf-like, large and prominent

18. Fr 6–8 mm wide; lf not papillate, not scabrous, base wedge-shaped, rounded, or truncate; (Otay Mt, s San Diego Co.) ..... ***A. otayensis*** (local endemic)

18' Fr 8–15 mm wide; lf papillate, scabrous, base truncate to ± lobed (Santa Catalina Island) ..... ***A. catalinae*** (local endemic) imm infl bracts fleshy, scoop-shaped

Baja California Key:

1. Pl with green to dull grey-green lvs, not glaucous white or grey

2. Pl with basal burls

3. Pl with bright, light green lvs and minute golden glands (Sierra San Pedro Martir)... ***A. patula* subsp. *patula***

3' Pl with dull green or grey-green lvs, glandular or not.

4. Pl with spheric fr, stones mostly fused(northern San Diego Co away from coast)... ***A. rainbowensis***[1]

4'. Pl with depressed-spheric fr, stones mostly separable

5. Pl lacking glandular hairs

6. Lvs green or grey-green, lacking reddish margin (coastal uplands in Baja)...***A. glandulosa* subsp *cushingiana***

6'. Lvs dark green, often with red margins (sw San Diego and n Baja coast)... ***A. glandulosa* subsp. *crassifolia***

6 Area covered by the **Key to the Baja California Region**

5'. Pl with long, often dense, glandular hairs (coastal mountains in San Diego and Baja) ... ***A. glandulosa* subsp. *glandulosa***

2'. Pl lacking basal burls

5. Lvs green, depressed-spheric fr with separable stones

6. Imm infl a dense, headlike raceme; non-glandular pubescence (interior mountains)... ***A. pungens***

6'. Imm infl a panicle, densely glandular pubescent (Punta Banda) ... ***A. glandulosa* subsp. *atumescens***

5'. Lvs grey-green (to glaucous), frs spheric with or without fused stones

7. Imm infl w leafy, pink, deciduous bracts (interior mountains of northern Baja)

8. Frs with stones fused ... ***A pringlei* subsp. *drupacea*[1]**

8'. Frs with stones separable ... ***A pringlei* subsp. *pringlei*[1]**

7'. Imm infl w appressed, persistent, scale-like bracts on spreading, thin branches; stones fused (gen < 1mm) (coastal hills and windward mountains, northern Baja) ... ***A. australis*[1]**

1'. Pl generally with glaucous white or grey lvs (some also grey-green)

9. Pl with basal burls

10. Fr depressed-spheric with mostly separable stones

11. Imm infl generally erect (foothills of southwestern Sierra Juarez Mountains) ... ***A. glandulosa* subsp. *erecta***

11'. Imm infl generally pendent

12. Twigs with dense glandular hairs (gen interiorcoast range from San Diego Co. south to San Tomas, Baja) ... ***A. glandulosa* subsp. *leucophylla***

12'. Twigs lacking glandular hairs

13. Twigs densely hairy (interior coast range from western San Diego south to Ojos Negros, Baja) ... ***A. glandulosa* subsp. *adamsii***

13'. Twigs lacking any hairs (vicinity of Cerro Bola Peak, interior coast range, Baja)... ***A. incognita***

10'. Fr spheric with fused stones

14. Spreading imm infl panicle mostly lax to erect, pedicels and ovary glandular hairy (divide west of San Antonio de las Minas, Baja) ... ***A. moranii***

14'. Imm infl not erect, pedicels and ovary lacking any hairs (Sierra San Pedro Martir south to Sierra San Borja Baja) ... ***A. peninsularis* subsp. *peninsularis***

9'. Pl lacking basal burls

15. Twigs generally lacking hairs

16. Spheric fr massive (10-15 mm wide), glandular sticky (widespread in the interior mountains from western San Diego to the Sierra San Borja, Baja) ... ***A. glauca***

16'. Spheric fr not massive (< 9 mm wide), smooth and not sticky

17. Imm infl with compact, stout panicle branches 5-10 long (interior Sierra Juarez mountains, Baja) ... ***A. peninsularis* subsp. *juarezensis***

17'. Imm infl with spreading, slender panicle branches -20 mm long (coastal hills and windward mountains, northern Baja)... ***A. australis*[2]**

15'. Twigs with dense glandular-tipped hairs

16. Imm infl with bright pink, lf-like bracts that are deciduous (interior mountains of northern Baja)

17. Spheric fr w stones fused ... ***A. pringlei* subsp. *drupacea*[2]**

17'. Spheric fr w stones separable ... ***A. pringlei* subsp. *pringlei*[2]**

16'. Imm infl with persistent lf-like bracts

19. Lf-like bracts 6-12 mm long, stiffly spreading (Otay Mountain, San Diego Co.)...***A. otayensis***

19'. Lf-like bracts 4-6 mm long, appressed (Cerro Bola, Baja) ... ***A. bolensis***

Outside the California Floristic Province Key:

7 Area covered by the key
Outside the California Floristic Province

1. Burls at base of main stem or where stems root, plants sprouting after fire; twig and/or infl axes and/or bracts with short golden glandular hairs;
burl often flat, obscure; (arid environments in eastern Oregon and eastern Calif, CFP) ..... ***A. patula* subsp. *patula***

1'. Burls 0; plants not sprouting after fire; twigs with various hairs, glandular or not

2. Lvs with stomata only on lower surface, surfaces gen different in color and/or hairiness, plants prostrate to mounding, lvs oblanceolate to obovate, occ narrowly elliptic, base wedge-shaped

3. Pedicels glabrous or hairy, not glandular; leaves obtuse, rounded, or apiculate at tip .... ***A. uva-ursi* subsp. *uva-ursi*** (widespread through-out North America and boreal regions in the Northern Hemisphere)

3'. Pedicels glandular hairy; leaf tips distinctly acute ...... ***A. uva-ursi* subsp. *cratericola*** (local endemic on the summits of volcanic peaks and limestone highlands of the Sierra de los Cuchumatanes in Guatemala).

2' Lvs with stomata on both surfaces, surfaces similar in color and/or hairiness

3. Pl prostrate to prostrate-mounded, 0.1–0.5(3) m; (mountains of Oreg, Wash) ..... ***A. nevadensis* subsp. *nevadensis***

3' Pl gen erect, occ mounded, (0.1)0.3–8 m

4. Imm infl bracts gen small, scale-like, deltate to awl-shaped, keeled or not,

5. Pl leaves green, imm infl and fruits not sticky glandular

6. Imm infl swollen at end, club-like, bracts recurved; twigs, infl non-glandular (Mexico n to Arizona, Texas);...***A. pungens***

6' Imm infl not swollen, infl and twigs with short golden glandular hairs (mnts in Oregon, Wash, Mont, Great Basin, Rocky Mountains to Ariz) .....***A. patula*[2] subsp. *patula***

5'. Pl leaves grey, imm infl and fruits sticky glandular (Oreg) ... ***A viscida* subsp. *viscida***

4' Imm infl bracts gen large and lf-like

7. Pl with grey leaves

8. Pl densely canescent (hairy)

9. Imm infl, pedicels, and ovary lacking glands (common, Oreg)... ***A. canescens* subsp. *canescens***

9'. Imm infl, pedicels, and ovary with glands (rare, Oreg) ... ***A canescens* subsp. *sonomensis***

7. Pl with green leaves

10. Lf-like bracts deep pink, glandular, deciduous after flwr. (Ariz) ....***A. pringlei* subsp. *pringlei***

10' Lf-like bracts green, glandular or not, persistent after flwr; (Pacific Coastal region from BC south to central Calif) ..... ***A. columbiana***

Manzanita Treatments

A. nevadensis in California's Sierra Nevada

Santa Cruz Manzanita

Arctostaphylos andersonii

FORM: tree-like, 2-5 m STEM: twig (and nascent inflorescence axis) densely tomentose or short-nonglandular-hairy and long-glandular-hairy LEAVES: overlapped; blade 4-7 cm, 1.5-2.5c m wide, oblong, boat-shaped, surfaces different, lacking stomata on upper leaf, glabrous, light-green, base lobed, clasping, tip acute INFLORESCENCE: panicle, 4-6-branched; nascent inflorescence pendent, axis 2-3 cm, > 1 mm wide; bracts 8-15 mm, +- leaf-like (occasionally reduced), lanceolate; pedicel 6-8 mm. FRUIT: depressed-globose; glandu- lar and sticky with separable stones HABITAT: open sites or forest edge, redwood or mixed-evergreen forest, occasionally in chaparral near coast; <800 m; Empire Grade, Santa Cruz Co. restricted to the southern Santa Cruz Mountains REMARKS: Only *A. andersonii* and *A. pajaroensis* have heart-shaped leaves with different leaf surfaces (lacking stomata above); *A. pajaroensis* has rough shredded bark compared to the smooth red bark of *A. andersonii*; one of 12 auriculate-leaved manzanitas in the Central Coast Range.

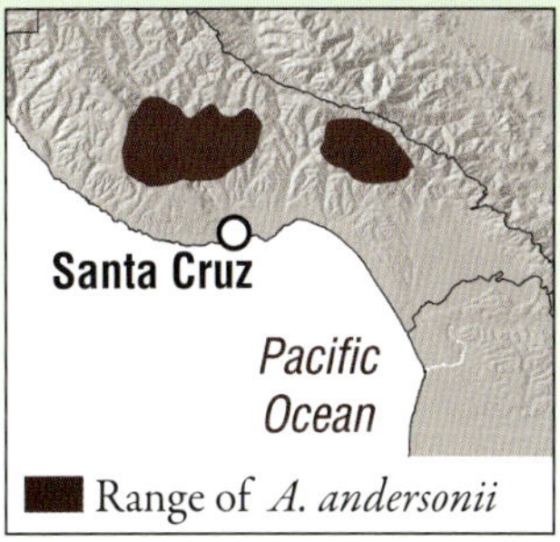

Mount Diablo Manzanita
Arctostaphylos auriculata

FORM: erect shrub, 1-4 m STEM: erect, twig white-tomentose LEAVES: overlapped; blade 1.5-4.5 cm, 1.5-3 cm wide, oblong- to round-ovate, white-gray, canescent, base lobed, clasping INFLORESCENCE: panicle, 3-5-branched; nascent inflorescence axis pendent, 1-1.5 cm, > 1 mm wide, tomentose; bracts 5-15 mm, leaf-like, ovate to lance-ovate, canescent; pedicel 4-10 mm. FRUIT: depressed-globose, 5-10 mm wide; stones free HABITAT: sandstone, upland chaparral, 150-650 m; Mount Diablo and Black Diamond Hills REMARKS: short pedicels and basal leaf lobes cause the leaves to clasp the branch; hairs are completely non-glandular throughout with a similar appearance to *A. gabilanensis* (fruit with seeds fused), *A. glutinosa* (fruit and branchlets that are glandular-sticky), and *A. luciana* of the Santa Lucia Mountains is especially beautiful in bloom with its white-glaucous leaves contrasting the deep rose-colored flowers

Range of *A. auriculata*

Baja Manzanita

Arctostaphylos australis

FORM: erect shrub, 1.5-2.5 m STEM: bark reddish and smooth LEAVES: 2-3.5 cm, elliptic, variable color from glaucous to grayish-green; smooth INFLORESCENCE: panicle, 5-6 branched, divergent in horizontally-spreading pattern, slender < 1 mm, buds well-spaced; bracts scale-like, 2-3 mm; creating 'beaded-string' appearance; pedicels 3-6 mm. FRUIT: 4-8 mm, ovoid HABITAT: coastal chaparral OBSERVATIONS: from Ensenada to San Vicente in coastal Baja California REMARKS: has been compared to *A. stanfordiana* (of the North Coast Range of California) because of its thread-like panicles with reduced floral bracts and spreading panicle branches; branchlets and inflorescence are completely smooth; leaf color is variable, from glaucous to green; commonly grows with the burl forming *A. glandulosa*.

Baker's Manzanita

Arctostaphylos bakeri subsp. *bakeri*

FORM: shrub, erect, 1-3 m STEM: erect; twigs glandular-bristly LEAVES: blade glandular-hairy, papillate, scabrous; 2-3.5 cm, erect, elliptic to ovate, dark-green, base generally wedge-shaped, flat INFLORESCENCE: nascent inflorescence axes 10-15 mm, >1 mm wide; bracts scale-like, 2-4 mm FRUIT: 8-10 mm wide, depressed-spheric; stones free HABITAT: serpentine chaparral near coast, 75-300 m; south of Guerneville in the Harrison Grade-Occidental area REMARKS: similar to *A. manzanita* with life form, multiple branched panicle, and scale-like bract; fruit is depressed-globose, smooth and with separable stones; *A. bakeri* subsp. *sublaevis* differs from subsp. *bakeri* by its tiny sessile glands beneath longer, nonglandular branchlet hairs. See range map next page.

Cedar's Manzanita

Arctostaphylos bakeri subsp. *sublaevis*

STEM: twig sparse- to dense-short-nonglandular-hairy with sessile glands LEAVES: blade appressed-puberulent; erect; blade 2.5-3 cm, 1.5-2 cm wide, elliptic to ovate, dark-green, base generally wedge-shaped, occasionally truncate to rounded, tip acute INFLORESCENCE: nascent inflorescence axes 15-20 mm, >1 mm wide; bracts scale-like, 2-4 mm FRUIT: 4-8 mm HABITAT: serpentine outcrops and ridges, 300-600 m; north to Healdsburg/Cazadero on serpentine outcrops REMARKS: grows in serpentine habitats in the "Cedars" (private) up through Cazadero/Healdsberg region.

Range of *A. bakeri* ssp. *bakeri*
Range of *A. bakeri* ssp. *sublaevis*

Cerro Bola Manzanita

Arctostaphylos bolensis

FORM: erect; 2-4 m; no burl LEAVES: glabrous, elliptic-ovate, glaucous-green STEM: glabrous INFLORESCENCE: showy panicle with 4-8 spreading branches with congested flowers; bracts bracts leafy, somewhat reduced 4-6 mm long FRUIT: 4-8 mm, ovoid; stone solid HABITAT: interior uplands, <400m OBSERVATIONS: Cerro Bola Mountain SW of Palm Valley REMARKS: related to *A. otayensis*, which differs in having stones with nutlets that are usually separable; unlike *A. otayensis*, which has erect panicles when developing, those of *A. bolensis* are pendant; range overlaps with *A. glandulosa* subsp. *adamsii* and *A. glandulosa* subsp. *leucophylla*, distinguished from these by taller, non-burl forming habit; flowers bloom prolifically, often covering a most of the plant's surface; limited distribution on Cerro Bola and reported from Cerro Italia.

Hoary Manzanita

Arctostaphylos canescens subsp. *canescens*

FORM: shrub, erect, 0.3–3 m STEM: twig short-canescent, nonglandular; twig canes- cent to white-tomentose LEAVES: erect; blade 2–5 cm, 1–3 cm wide, round-ovate, ovate, or elliptic, white-gray, canescent, base rounded to wedge-shaped INFLORESCENCE: raceme to panicle, 1-3-branched; nascent inflorescence pendent, bell-shaped, axis 1-2 cm, >1 mm wide, +- obscured, +- canescent to white-tomentose; bracts 6-20 mm, leaf-like, widely lanceolate, canescent; pedicel 5--9 mm, densely hairy FRUIT: densely-hairy early fruit, glabrous in age; 5–10 mm wide, depressed-spheric; stones free HABITAT: chaparral, open forest; 200–2000 m; widespread in North Coast Range, found on Mt. Tam, Mt. St. Helena, and in Mayacamas Mts.; see range map next page REMARKS: three auriculate-clasping species (*A. auriculata, A. luciana,* and *A. glutinosa*) have similar foliage vestiture as *A. canescens* but differ mainly in leaf shape or in having glandular hairs on the inflorescence, ovaries, and fruit; hybridizes with *A. viscida* in the northern part of the range where the distribution of these two species overlaps extensively; closely related to *A. malloryi* (Redding area), *A. silvicola* (Santa Cruz Mtns.), and *A. obispoensis* (Santa Lucia Mtns.): light colors of leaves caused by canescent hairs, hairs fall off with age and leaves stay whitish because of glaucous wax covering.

Range of *Arctostaphylos canescens* subsp. *sonomensis*
Arctostaphylos canescens subsp. *canescens*

Sonoma Manzanita

Arctostaphylos canescens subsp. *sonomensis*

STEM: twig canescent, glandular INFLORESCENCE: densely glandular hairy FRUIT: glandular-hairy HABITAT: chaparral, open forests; 60–1700 m OBSERVATIONS: scattered from Santa Cruz Co. north to Humboldt Co. REMARKS: given its widespread occurrence in mixed stands with non-glandular *A. canescens*, it is being reconsidered as a special status taxon and may ultimately be classified as a genetic variant within the *A. canescens* species complex

Catalina Island Manzanita

Arctostaphylos catalinae

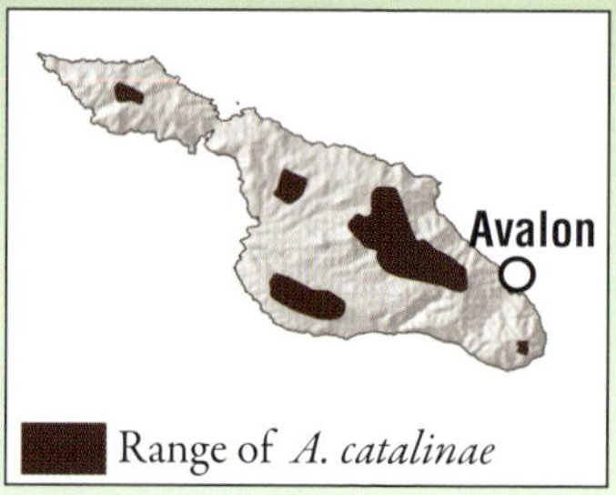

FORM: shrub, small tree, erect, 2–5 m STEM: twig (and inflorescence bracts) densely glandular- and long-white-nonglandular-hairy LEAVES: over-lapped; blade 2–5 cm, 1.5–3 cm wide, lance-ovate to elliptic, green-glaucous, glandular- bristly, papillate, scabrous, base truncate to ±lobed, tip acute INFLORESCENCE: panicle, 4-10-branched; nascent inflorescence pendent, axis 2-3 cm, > 1 mm wide; bracts generally overlapped in distal 1/2 of inflorescence, 6-10(15) mm, leaf-like, lanceolate to narrowly ovate, acute, green, persistent; pedicel 2-5 mm, tomentose FRUIT: 8-15 mm wide, spheric-globose; stones free HABITAT: volcanic outcrops, ridges, maritime chaparral; 100-600 m; restricted to Catalina Island REMARKS: large, generally erect non-burl forming manzanita with glaucous-green leaves restricted to Catalina Island; *A. catalinae* and *A. confertiflora* (from Santa Rosa Island) both have short glandular hairs with long white bristles on their branchlets; *A.catalinae* differs from *A. confertiflora* primarily in its habit (*A. confertiflora* is typically mound forming rather than erect), more glaucous foliage (*A. confertiflora* is light green), more open inflorescence, different colored fruit, and leaves that are not as rounded; only species on the Santa Catalina Island.

Columbia Manzanita

Arctostaphylos columbiana

FORM: shrub, small tree, erect, 2–5m STEM: twigs densely tomentose, generally with long white bristles; glandular or not LEAVES: spreading; blade 4–6cm, 2–3cm wide, lanceolate to ovate, flat, dark green, dull, sparsely glandular-hairy, ±papillate, scabrous, base wedge-shaped to ±rounded, tip acute INFLORESCENCE: panicle, 3--8-branched; nascent inflorescence pendent, axis 1.5--2.5 cm, > 1 mm wide; bracts leaf-like, green, persistent, stiff, densely short- and long-white-nonglandular-hairy, occasionally glandular-hairy; pedicel 2--4 mm, glandular-hairy.FRUIT:8–11mm wide, depressed-spheric, sparsely nonglandular-hairy; stones free HABITAT: rocky coastal uplands, maritime chaparral, conifer forest edge; <800m; north-coastal California to British Columbia, f. tracyii without hairs in Patricks Point State Park REMARKS: long, white bristles on its branchlets and inflorescence; hybridizes with A. uva-ursi where ranges overlap; hairs absent in form tracyii in northern California; one of the few manzanita species that extend far beyond California.

Santa Rosa Island Manzanita

Arctostaphylos confertiflora

FORM: shrub, prostrate to mounded or erect; 0.1-2m STEM: twigs densely tomentose and with long, white, gland-tipped bristles LEAVES: ovate to elliptic, light-green, dull, ±glandular-hairy near base; 2-4 cm (1.5-3.5 cm wide) INFLORESCENCE: panicle, 3-5-branched; nascent inflorescence pendent, axis 1.5-2 cm, >1 mm wide; bracts +- densely overlapped near inflorescence tip, 8-18 mm, leaf-like, ovate to oblanceolate, green, densely short- and long-white-glandular-hairy, persistent; pedicel 3-5 mm, glandular-hairy FRUIT: 8–11 mm wide, sparsely hairy HABITAT: sandstone outcrops, maritime island chaparral, closed-cone conifer forests; of conservation concern; <500 m; Santa Rosa Island, in and around the Torrey pine (*Pinus torreyana* subsp. *insularis*) forest often on calcareous soils REMARKS: grows on Santa Rosa Island with *A. crustacea* subsp. *subcordata* and subsp. *insulicola*, which differ in having bifacial leaves and basal burls; similar to *A. catalinae* but growth form is a mounding version of the larger, erect, Catalina manzanita; species is of conservation concern.

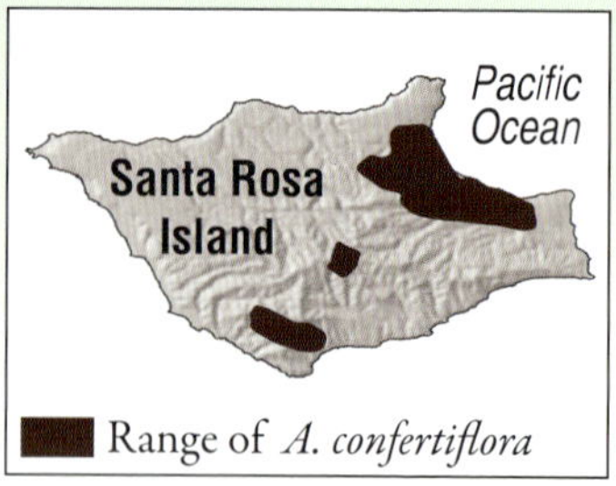

Brittleleaf Manzanita

Arctostaphylos crustacea subsp. *crustacea*

FORM: erect, 1-3 m; burl prominent STEM : twigs with dense short hairs and long spreading hairs, generally not glandular; older stems with smooth, red bark LEAVES: spreading, 1.5-2.5 cm wide, 2-5 cm long, dark-green above, light-green below, glabrous or moderately; nonglandular-hairy, base truncate to ± lobed; no stomata on upper surface INFLORESCENCE: panicle, 2--8-branched; nascent inflorescence pendent, axis 1--2.5 cm, >1 mm wide, rachis hairs both short and long, non-glandular, ovary non-glandular hairy; lower bracts leaf-like, upper often scale-like; bracts 3--8 mm, scale-like and awl-shaped to deltate or leaf-like and lanceolate, glabrous to hairy; pedicel 2--5 mm, sparsely nonglandular-hairy. FRUIT: 6-10mm wide, depressed-spheric, glabrous to sparsely nonglandular-hairy; stones free HABITAT: generally within or adjacent to the fog belt; A. crustacea is the most widespread subspecies, ranging from Mount George (Napa) through East Bay Hills, Santa Cruz Mountains, Channel Islands. REMARKS: *A. crustacea* was formerly treated as a subspecies of *A. tomentosa* (persistent shreddy bark), the two burl-forming species are now separated due to their different bark characteristics; the species is divided into six subspecies with differences mainly having to do with their hairs and glands on branchlets, inflorescence, ovaries, and fruit.

Crinite Manzanita

A. crustacea subsp. *crinita*

LEAVES: Similar to subsp. *crustacea* except densely tomentose on lower leaf, occasionally hairy on upper leaf HABITAT: chaparral, conifer forest, <700m Central Coast; western Santa Cruz Mountains; Fort Ord Hills, Del Monte Forest, San Luis REMARKS: Mostly southwestern Santa Cruz Mountains, sometimes individuals appear whitecanescent, as on Ice Cream Grade; a more densely hairy version of subsp. *crustacea*.

Range of *Arctostaphylos crustacea*

subsp. *rosei* at Bixby Bridge ▲

Eastwood's Manzanita

A. crustacea subsp. *eastwoodiana*

STEM: twigs with dense, short non-glandular hairs; lacking long hairs, LEAF: lacking hairs on the lower surface of the leaves; glabrous INFLORESCENCE: unlike subsp. *rosei*, pedicels and ovary are glabrous HABITAT: chaparral, closed-cone conifer forest; <650 m, s Central Coast/Outer South Coast Ranges (nw Santa Barbara Co.) REMARKS: very limited range.

Island Manzanita

A. crustacea subsp. *insulicola*

STEM: twigs with dense, short non-glandular hairs; lacking long hairs LEAF: lower surface densely non-glandular hairy INFLORESCENCE: rachis and ovary short non-glandulary hairy HABITAT: chaparral, conifer forest; 5–700 m, (Santa Cruz, Santa Rosa islands) REMARKS: similar to *A. tomentosa* subsp. *tomentosa* except with smooth, red bark.

Rose's Manzanita

A. crustacea subsp. *rosei*

STEM: twigs with dense, short non-glandular hairs; lacking long hairs LEAF: glabrous INFLORESCENCE: ovary densely hairy HABITAT: chaparral, conifer forest, coastal bluffs; <500 m, Central Coast (Lake Merced, San Francisco Co.; coastal Big Sur, northern Monterey Co.) REMARKS: extremely rare in San Francisco and limited along Big Sur coast.

Santa Cruz Island Manzanita

A. crustacea subsp. *subcordata*

HABITAT: chaparral, conifer forest; 100-730 m, (Santa Cruz, Santa Rosa islands) IDENTIFICATION: distinguished from non-glandular *A. crustacea* subsp. *insulicola* by its dense, long, glandular hairs on branchlets and inflorescence; most widespread and abundant manzanita of the Channel Islands; occasional glandular individuals can be found in other populations of *A. crustacea* (e.g., San Bruno Mountain, San Mateo County).

Arroyo De La Cruz Manzanita

Arctostaphylos cruzensis

FORM: shrub, prostrate, 0.1-1 m STEM: twig (and nascent inflorescence axis, bract, pedicel) sparsely nonglandular-hairy LEAVES: overlapped, bright green, tomentose, glabrous in age, smooth, base lobed, clasping INFLORESCENCE: panicle, 1-3-branched; nascent inflorescence pendent, bell-shaped, crowded, hidden by large lower bracts, axis 0.5-1.5 cm, >1 mm wide; bracts 5-15 mm, leaf-like, lanceolate to lance-ovate; pedicel 4-5 mm.

FRUIT: 8-10 mm wide, depressed- spheric, smooth without hairs; stones free HABITAT: sandy bluffs, maritime chaparral, coastal prairie; <150 m; restricted to Arroyo de la Cruz on the Hearst Ranch and Arroyo de los Chinos near Point Piedras Blancas; grows with another prostrate manzanita, *A. hookeri* subsp. *hearstiorum*; *A. tomentosa* subsp. *tomentosa* is also present; closely related to other auriculate-leaved manzanitas such as *A. osoensis* and *A. morroensis* REMARKS: only auriculate-leaved manzanita that is prostrate or low mounding in habit besides *Arctostaphylos imbricata*

Vine Hill Manzanita

Arctostaphylos densiflora

FORM: shrub, mounded to erect, 1 m STEM: twig (and nascent inflorescence axis) sparsely short-nonglandular-hairy LEAVES: erect; blade 1-2.5 cm. 0.5-1.5 cm wide, elliptic to narrowly lanceolate, bright-green, shiny; base wedge-shaped to obtuse, tip acute INFLORESCENCE: panicle, 3-5-branched; nascent inflorescence pendent, axis 1-1.5 cm, <1 mm wide; bracts 1.5-3 mm, > buds, scale-like, awl-shaped to deltate, glabrous; pedicel 4-5 mm, glabrous. FRUIT: ±spheric, glabrous; stones free HABITAT: marine acid sandy-clay soils, maritime chaparral, ±100m; Vine Hill near Forestville, Sonoma Co. REMARKS: local endemic in Sonoma County whose native habitat has been reduced to a small remnant of its former distribution; now occurs on about one acre in the Vine Hill Preserve; appears to be most closely related to *A. stanfordiana*; *A. manzanita* subsp. *manzanita* individuals grow near the Vine Hill Preserve and the two appear to hybridize

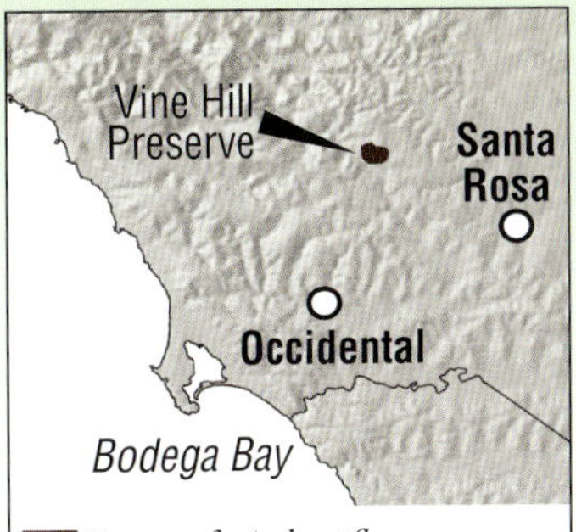

Range of *A. densiflora*

Little Sur Manzanita

Arctostaphylos edmundsii

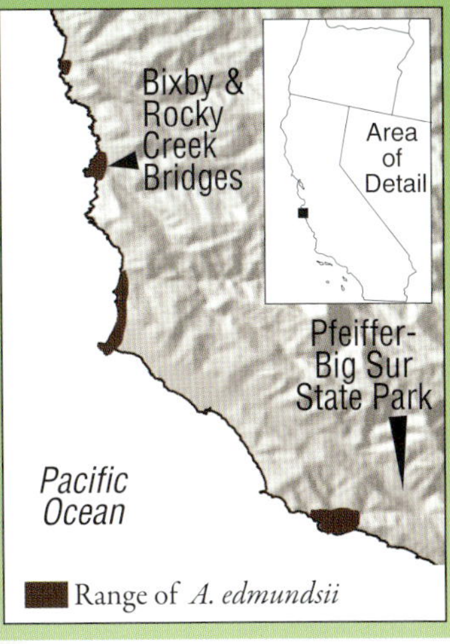

FORM: shrub, prostrate to mounded, 0.2-1.5 m STEM: twig (and nascent inflorescence axis) sparsely short-nonglandular-hairy LEAVES: spreading; blade 1–2.5 cm, 1–1.5 cm wide, elliptic to round-ovate, dark green, shiny, base truncate to lobed, tip abruptly soft-pointed INFLORESCENCE: panicle, 2-5-branched; nascent inflorescence pendent, axis 0.5-1 cm, > 1 mm wide; bracts 3-5 mm, scale-like, deltate to awl-shaped, (and pedicel) glabrous or sparsely nonglandular-hairy; pedicel 3-4 mm. FRUIT: ±spheric, glabrous; stones free HABITAT: sandy terraces, bluffs, maritime chaparral; <100 m; Pfeiffer Beach State Park, Garrapata State Park REMARKS: similar to *A. uva-ursi* which co-occurs in the same area; however, *A. edmundsii* has more broad leaves with truncate to lobed bases with isofacial leaves, *A. uva-ursi* leaves are dark green above and light below with bifacial leaves

Franciscan Manzanita
Arctostaphylos franciscana

FORM: prostrate to prostrate-mounded, 0.2-1.5 m STEM: twig gray-tomentose, nonglandular LEAVES: spreading; blade 1.5-2 cm, 0.5-1 cm wide, glabrous in age, smooth, bright green, base wedge- shaped, tip acute INFLORESCENCE: panicle, 1-3-branched; nascent inflorescence pendent, axis 0.5-1 cm, >1 mm wide, gray-nonglandular-hairy, bracts near base leaf-like, above scale-like, 3-4 mm, awl-shaped, concave when fresh, fleshy, nonglandular; pedicel 2-6 mm, glabrous. FRUIT: 6-8 mm wide, depressed-spheric, glabrous; stones free HABITAT: serpentine outcrops, chaparral; <300 m OBSERVATIONS: one plant in wild, recently discovered on the Presidio; planted in East Bay Botanic Gardens IDENTIFICATION: *A. franciscana* (mounding) and *A. montana* subsp. *ravenii* (prostrate) may have grown together in San Francisco in the past REMARKS: thought to be extinct in the wild since its last population at Laurel Hill Cemetery was destroyed in the 1940's, a single wild plant discovered in 2009 on a serpentine outcrop along Doyle Drive near the Golden Gate Bridge, translocated to an undisclosed location in the Presidio.

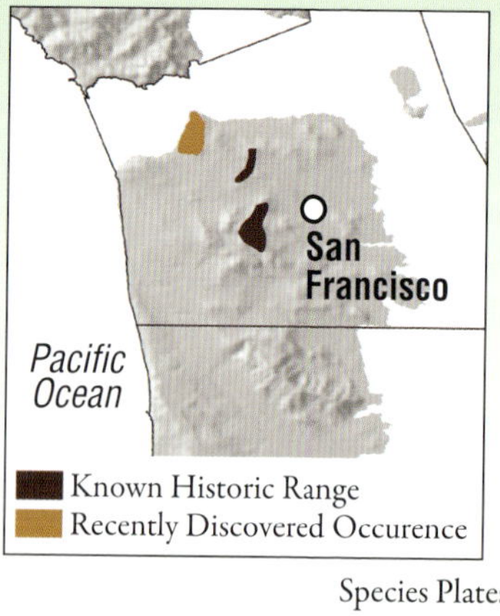

Gabilan Manzanita

Arctostaphylos gabilanensis

FORM: erect, 1-5m STEM: twig (and lower inflorescence bracts) sparsely short-nonglandular-hairy, with sessile glands LEAVES: overlapped; blade 1.5-3.5 cm, 1.1-2.4 cm wide, ovate, gray-green, glabrous or generally short-nonglandular-hairy, base lobed, clasping, tip obtuse INFLORESCENCE: panicle, 1-4-branched; nascent inflorescence pendent, axis 1-3 cm, >1 mm wide; bracts 3-5 mm, fleshy, deltate, nonglandular-hairy, lower leaf-like; pedicel 8-10 mm, glabrous or nonglandular-hairy. FRUIT: 10-15 mm wide, spheric, glabrous; stones fused HABITAT: open granitic outcrops, chaparral, coulter-pine/chaparral woodland; <710 m: two populations are known, the smallest near Gloria road and larger growing southeast of Fremont Peak REMARKS: fruit is smooth and non-glandular in contrast to the sticky, viscid fruits of *A. glauca*; described in 2004, this species escaped discovery because its entire known distribution is on private ranch property in the Gabilan Range.

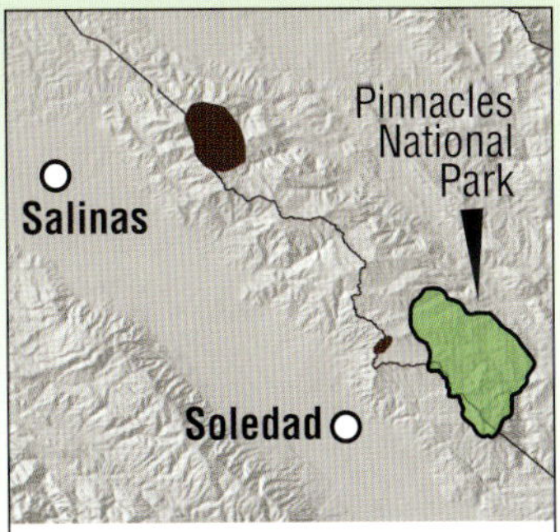

Range of *A. gabilanensis*

Eastwood's Manzanita

A. glandulosa subsp. *glandulosa*

FORM: shrub, erect, 1-4 m; burl hemispheric STEM: twig (and nascent inflorescence axis, pedicel) glandular-hairy LEAVES: ascending or erect; blade 2-4.5 cm, 1–2.5 cm wide, elliptic to ovate, bright-green to gray-green, shiny or dull, glandular-puberulent to hairy, papillate, base wedge-shaped to rounded INFLORESCENCE: panicle, 3-6-branched; nascent inflorescence pendent, +- crowded, hairy, glandular-hairy; axis 1-3 cm, > 1 mm wide; pedicel 3-10 mm. FLOWERS: ovary short glandular hairy FRUIT: 6-10 mm wide, depressed-spheric, glandular-hairy; stones generally free HABITAT: chaparral, conifer forest; 30-1900 m; Central and Southern Oregon, Klamath Range, North Coast Range, San Francisco Bay Area, Central Coast Range, South Coast Range, Peninsular Range; northern Baja California REMARKS: most widespread subspecies characterized by its glandular-hairy parts; varies from green to gray-green throughout its extensive range.

Adam's Manzanita

A. glandulosa subsp. *adamsii*

STEM: twigs short non-glandular harry LEAVES: non-glandular, intensely white-glaucous INFLORESCENCE: rachis, pedicels, and ovary densely non-glandular hairy HABITAT: chaparral, conifer forest; 1500-2200 m southwestern California; northern Baja California REMARKS similar range and appearance as subsp. *leucophylla* except lacking glandular hairs

Punta Banda Manzanita

A. glandulosa subsp. *atumescens*

FORM: Lacks a burl, otherwise same as subsp. *glandulosa* HABITAT: coastal chaparral, <500 m, southern end of Ensenada Bay in Baja California REMARKS: known only from Punta Banda; a 2008 fire may have eliminated adult shrubs but they should be recruiting; subspecies needs further study.

Del Mar Manzanita

A. glandulosa subsp. *crassifolia*

STEM: Non-glandular with short-white hairs LEAVES: green leaves variable, sometimes gray leaved, often with reddish edges, usually elliptical INFLORESCENCE: axis, bract, pedicel, ovary densely white-nonglandular-hairy, bracts occasionally leaf-like; FRUIT: highly flattened fruit HABITAT: maritime chaparral; <100 m, from Torrey Pines State Park to Cabo Colonet, Baja California; REMARKS: listed as an endangered species.

Cushing's Manzanita

A. glandulosa subsp. *cushingiana*

STEM and INFLORESCENCE:: non-glandular with dense, short white hairs FRUIT: non-glandular HABITAT: chaparral, conifer forest; 50–1900 m; Klamath Ranges, North Coast Range, San Francisco Bay Area, South Coast Ranges, Peninsular Ranges; northern Baja California REMARKS: found mixed with subsp. *glandulosa* from near the Oregon border to San Vicente in Baja California; occasionally dominant and sometimes pure stands.

Pedregosos Manzanita

A. glandulosa subsp. *erecta*

INFLORESCENCE: branchlets densely hairy, non-glandular; distinguished by its erect immature inflorescences, which form in summer after the flowers fade and remains fully formed for over half the year HABITAT: chaparral, interior Coast Range, 200-500 m; east of Ensenada on Cerros Pedregosos REMARKS; no other subspecies of *A. glandulosa* has consistently erect panicles; needs further study.

San Gabriel Manzanita

A. glandulosa subsp. *gabrielensis*

STEMS and INFLORESCENCE: non-glandular short hairy FRUIT: depressed spheric to spheric like *A. glandulosa* but with stones fused into a pumpkin-shaped unit HABITAT: chaparral on granitic soils; 950-2000 m; interior, higher elevations of the San Gabriel Mountains, also west in the Sierra Madre Mountains of Santa Barbara County REMARKS: only subspecies with fused stones, a trait also characteristic of nearby populations of *A. parryana*.

Howell's Manzanita

A. glandulosa subsp. *howellii*

STEM: densely short non-glandular hairy INFLORESCENCE: glandular-hairy HABITAT: chaparral, conifer forest; 400-1800 m, outer central Coast Ranges, restricted to the Santa Lucia Mountains where it associates with *A. hooveri* and *A. obispoensis*; REMARKS: similar to subsp. *cushingiana* with branchlets that have only sparse-to-dense non-glandular short hairs but differs in having a glandular-hairy immature inflorescence, similar to subsp. *glandulosa*.

Whiteleafed Manzanita

A. glandulosa subsp. *leucophylla*

STEM: long glandular hairy LEAVES: glaucous white, glandular hairy INFLORESCENCE: densely white non-glandular hairy rachis, pedicel, and ovary HABITAT: chaparral, conifer forest; 200-1930 m; w San Bernardino Mountains, Peninsular Ranges; northern Baja California REMARKS: similar to subsp. *adamsii*, subsp. *leucophylla* has intensely white-glaucous leaves, but differs in having glandular hairs; both subspecies are mainly found from San Diego County south into northern Baja California and may occur in mixed populations.

Transverse Range Manzanita

A. glandulosa subsp. *mollis*

STEM: non-glandular, short white hairs with long, wavy setose hairs LEAVES: shiny, smooth, green to gray green INFLORESCENCE: densely non-glandular hairy rachis, pedicel, and ovary HABITAT: chaparral, conifer forest; 60-2000 m, Transverse Range, Peninsular Range, and northern Baja REMARKS: pubescence pattern for stems, rachis, and ovary similar to *A. crustacea* but leaves place subspecies within *A. glandulosa*.

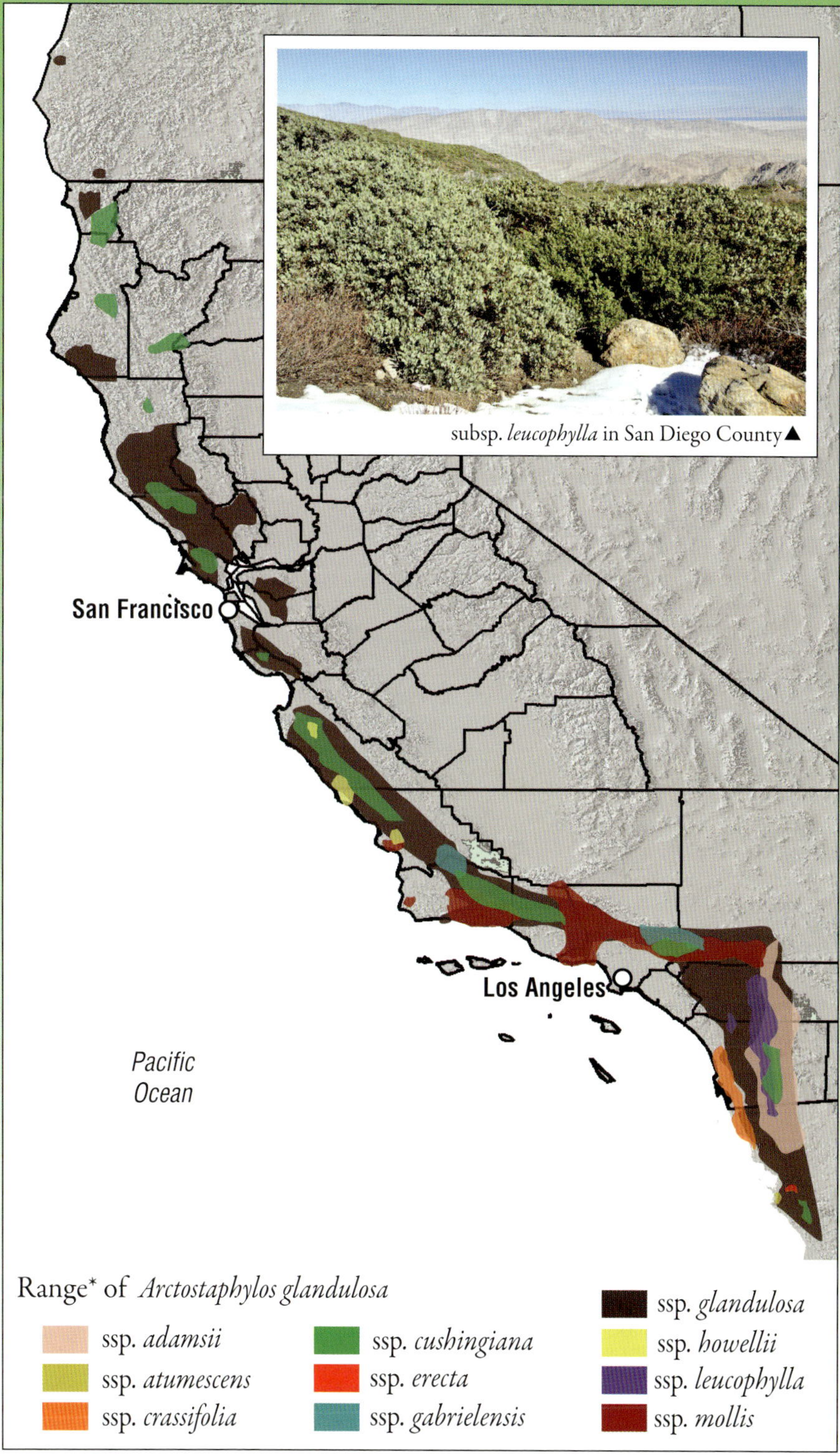

subsp. *leucophylla* in San Diego County ▲

Bigberry Manzanita

Arctostaphylos glauca

FORM: shrub, small tree, erect, 1-8 m STEM: twig usually glabrous or sometimes having nonglandular or glandular hairs LEAVES: erect; blade 2.5-5 cm, 2-4 cm wide, oblong-ovate to ±round, white-glaucous, dull, glabrous, base rounded, truncate, tip abruptly soft-pointed FLOWERS: ovary densely glandular-sticky INFLORESCENCE: panicle, 4-8-branched; nascent inflorescence pendent, axis 2-3 cm, >1 mm wide, glabrous, occasionally short-nonglandular or glandular-hairy; bracts spreading, 3-6 mm, scale-like, deltate, scoop-shaped, glabrous; pedicel 8-10 mm, glandular-hairy FRUIT: 10-15 mm wide, spheric-globose; stones fused into single structure; quite large and glandular-sticky HABITAT: rocky slopes, chaparral, woodland; <2200 m; representative sites at Mount Diablo, western Mojave, and Sierra de San Pedro Mártir N.P. REMARKS: largest fruit of all manzanita species; white-glaucous leaves are similar to those of *A. viscida* in the Sierra Nevada and North Coast Ranges but different in spreading, scoop-shaped bracts instead of tightly appressed bracts; one of the largest and most common manzanitas of southern California and Baja California.

Schreiber's Manzanita

Arctostaphylos glutinosa

FORM: shrub, erect, 1-2 m STEM: twig sparsely to densely long-glandular-hairy LEAVES: overlapped; blade 2-5 cm, 1-3 cm wide, oblong to oblong-ovate, glaucous, dull, canescent, base lobed, clasping, tip acute FLOWERS: ovary densely long-white hairs, glandular INFLORESCENCE: panicle, 2-4-branched; nascent inflorescence pendent, axis 1.5-2.5 cm, >1 mm wide, sparsely to densely soft-glandular-hairy; bracts 5-15 mm, leaf-like, lance-oblong, gray-canescent; pedicel 5-8 mm, glandular-hairy. FRUIT: 7-14 mm wide, depressed-spheric, glandular-sticky; stones free HABITAT: siliceous shale outcrops, chaparral, knobcone-pine woodland; 180-650 m; Lockheed Chalks, Santa Cruz Mts. (not open to the public) REMARKS: grows with five other manzanita species in a "hot spot" for manzanita diversity: *A. ohloneana, A. crustacea* subsp. *crinita, A. andersonii, A. sensitiva,* and *A. silvicola.*

Gasquet Manzanita

Arctostaphylos hispidula

FORM: shrub, erect, 1-3 m STEM: twig fine-glandular-hairy LEAVES: erect; blade 1.5-3 cm, 0.5-1.5 cm wide, elliptic to oblanceolate, dull green, glandular-hairy, base wedge-shaped, tip acute, margin entire, flat INFLORESCENCE: panicle, 3-6-branched; nascent inflorescence spreading to ascending, axis 1-2 cm, <1 mm wide, glandular-hairy; bracts 2-4 mm, scale-like, awl-shaped, glandular-hairy; pedicel 3-5 mm, glabrous. FRUIT: 5-7 mm wide, ± spheric, glabrous; stones free HABITAT: rocky serpentine soils or sandstone, interior chaparral, open woodland; 100-1250 m; for example, Smith River NRA on the High Divide REMARKS: co-occurs with *A. nortensis* and *A. nevadensis* subsp. *knightii*; appears closely related to *A. stanfordiana* but has fine glandular hairs on the branchlets and dull green leaves with a rough, bristly feel; also found in Sonoma County in serpentine areas along Rockpile Road above Lake Sonoma and in the Cedars (not shown on range map).

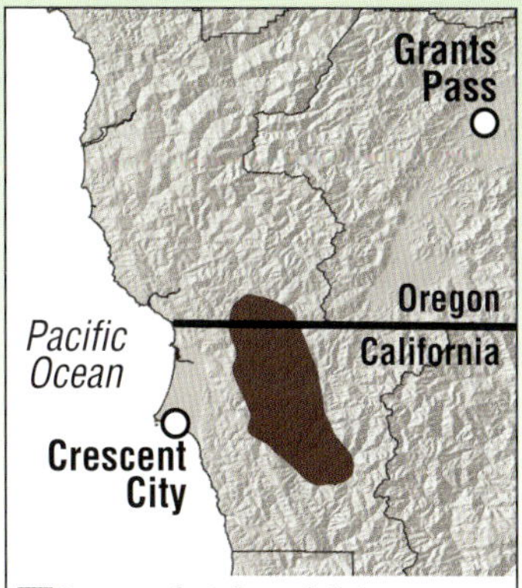

Range of *A. hispidula* (see remarks)

Hearst's Manzanita

Arctostaphylos hookeri subsp. *hearstiorum*

FORM: shrub, prostrate, 0.1-0.2 m LEAVES: blade 0.8-1.2 cm, 4-7 mm wide, ±elliptic, diamond-shaped FRUIT: 3-4 mm wide HABITAT: coastal prairie; 10-50 m OBSERVATIONS: Arroyo de la Cruz, part of the vast Hearst Ranch near San Simeon in San Luis Obispo County REMARKS: subsp. *hearstiorum* differs from subsp. *hookeri* in its prostrate nature and diamond shaped leaves as opposed to upright growth and elliptic leaves of subsp. *hookeri*; associates with *A. cruzensis* which is a mounding shrub but has auriculate-clasping blue-green colored leaves; grows on property that is private and permission is required to visit the site.

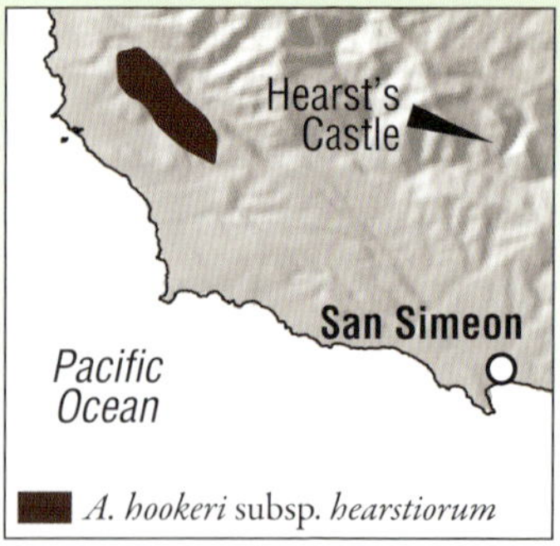

Hooker's Manzanita

Arctostaphylos hookeri subsp. *hookeri*

FORM: shrub, mounded to erect, 0.5-2 m LEAVES: petiole 4-8 mm; blade 2-3 cm, 1-1.5 cm wide, lanceolate to widely elliptic INFLORESCENCE: +- raceme, 0-1-branched; nascent inflorescence pendent, spheric or cup-like, dark, axis 0.3-1 cm, >1 mm wide, crook-necked; bracts 2-5 mm, +- spreading, scale-like, deltate, glabrous; pedicel 2-6 mm, glabrous FRUIT: depressed-spheric; 4-8 mm wide HABITAT: chaparral, closed-cone pine forest; <200 m; for example, Point Lobos State Reserve; Morse Botanical Reserve at Huckleberry Hill; REMARKS: growing with *A. pajaroensis* and *A. crustacea* at Manzanita Park in Prunedale; A bright-green, small-leaved manzanita with predominantly white flowers; either as an erect shrub (to 2 m) or a mound-forming bush or low prostrate shrub (especially in the Monterey pine forest understory).

Range of *A. hookeri* ssp. *hookeri*

Hoover Manzanita

Arctostaphylos hooveri

FORM: shrub, small tree, 2-8m STEM: twig (and nascent inflorescence axis) densely long-glandular-hairy LEAVES: blade 4-6 cm, 2-3 cm wide, oblong to ovate, boat-shaped, glaucous, dull, glandular-sticky-hairy, base lobed, not clasping, tip acute, margin entire, flat INFLORESCENCE: panicle, 4-6-branched; nascent inflorescence pendent, axis 1.5-2.5 cm, > 1 mm wide; bracts 8-20 mm, leaf-like, lanceolate, acuminate, glandular-hairy; pedicel 8-15 mm, glandular-hairy FRUIT: 10-15 mm wide, depressed-spheric, glandular-hairy, sticky; stones free HABITAT: Rocky slopes, upland chaparral, open ponderosa-pine forest near coast; 450-1100 m; for example, the Cone Peak area of the Santa Lucia Mountains REMARKS: large, robust species with thick trunks and branches occurring in the western Santa Lucia Mountains, leaves distinct in being dull-gray and glandular-sticky at first, becoming smooth with age; similar to *A. regismontana* of the Santa Cruz Mountains, but ranges do not overlap; sometimes associates with ponderosa pine.

San Bruno Mountain Manzanita

Arctostaphylos imbricata

FORM: shrub, prostrate, 0.1-1 m STEM: twig densely long-glandular-hairy LEAVES: overlapped; blade 2.5-4 cm, 2-3 cm wide, round to round-ovate, light-green, sparsely glandular-hairy, base lobed, clasping, tip abruptly soft-pointed, flat INFLORESCENCE: panicle, 3-5-branched; nascent inflorescence pendent, axis 0.5-1 cm, >1 mm wide, long-glandular-hairy; bracts 5-10 mm, leaf-like, ovate, acute, glandular-hairy; pedicel 3-5 mm, glandular-hairy. FRUIT: 6-7 mm wide, depressed-spheric, glandular-hairy, sticky; stones free HABITAT: sandstone outcrops, chaparral; 200-400 m; San Bruno Mountain State Park REMARKS: restricted to San Bruno Mountain in northern San Mateo County, *A. imbricata* is a prostrate, mat-forming manzanita with auriculate-clasping leaves; related to *A. montaraensis* of which a few also grow at Pacific Rock on San Bruno Mountain; associates with three other species of manzanita: *A. paciftca, A. uva-ursi; and A. montaraensis.*

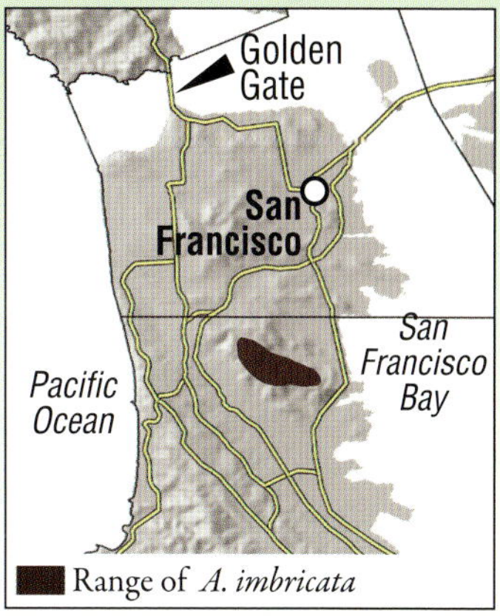

Incognito Manzanita

Arctostaphylos incognita

FORM: shrub, 1-4m; large globose burl LEAVES: with equal number of stomata on both sides, blades 3.5-4 cm long, 2.5-3 cm wide, elliptic, glabrous and heavily white-glaucous, base acute to rounded INFLORESCENCE: panicle with 5-6 branches; bracts 1-2 mm, overlapping, deltoid, long-pointed, keeled, rachis glabrous FRUIT: depressed-spheric, 5-6 mm long, 8-9 mm wide, glabrous; red; mesocarp thick and mealy, nutlets separable HABITAT: rhyolite soils; 300-700 m; in the hills above Carmen Serdán, about 40km south of Tecate in Baja California REMARKS: appears to be related to *A. glandulosa* and possibly *A. peninsularis*; similar to *A. peninsularis* in being completely glabrous with no hairs or glands on its branchlets and inflorescence, distinct in fruits being depressed spheric (wider than long) with separable nutlets; epithet was inspired by the fact that, although populations are easily visible from one of two major roads, species went undescribed until 1997.

Santa Cruz Island Manzanita

Arctostaphylos insularis

FORM: shrub, erect, 2-5 m STEM: twig (and nascent inflorescence axis) glabrous (sparsely short-nonglandular- or sparsely glandular-hairy) LEAVES: spreading or erect; blade 2.5-4.5 cm, 1-3 cm wide, oblong-elliptic, cupped, bright green, shiny, base rounded, tip obtuse; stomata on bottom only INFLORESCENCE: panicle, 5-10-branched; nascent inflorescence pendent, axis 2-3 cm, >1 mm wide; bracts 1.5-3 mm, scale-like, deltate, glandular-hairy; pedicel 6-9 mm, glandular-hairy FRUIT: 8-15 mm wide, depressed-spheric; stones free HABITAT: rocky slopes, chaparral, woodland; <400 m; Prisoner's Harbor and Pelican Bay, on the north side of Santa Cruz Island REMARKS: bark with frequent white blushes on the older large stems; distinct from the other Channel Island species with bright-green, shiny leaves, stomata restricted to the bottoms of the leaves although color and pubescence not markedly different, and one of the showiest clusters of white flowers of any manzanita species; associates with *A. crustacea* subsp. *subcordata*, subsp. *insulicola*, and *A. viridissima*.

Klamath Manzanita

Arctostaphylos klamathensis

FORM: shrub, prostrate, 0.1-0.5 m STEM: twig (and nascent inflorescence axis, bract, pedicel, ovary) glandular-hairy LEAVES: erect; blade 1-3.5 cm, 0.5-2.5 cm wide, obovate to widely elliptic, glaucous, dull, midvein sparsely glandular-hairy, base wedge-shaped or obtuse, tip obtuse, flat INFLORESCENCE: +- raceme, 0-1-branched; nascent inflorescence pendent, axis 0.5-1 cm, > 1 mm wide, crook-necked; bracts 2-5 mm, scale-like, awl-shaped; pedicel 3-6 mm FRUIT: 6-7 mm wide, spheric, glabrous; stones fused HABITAT: Rocky outcrops, slopes, subalpine forest; 1600-2000 m, eastern Klamath Ranges; for example, Scott Mtn. Summit Campground, Slate Mtn, Cedar Basin Research Natural Area REMARKS: although prostrate to decumbent, *A. klamathensis* is usually taller than *A. nevadensis* with which it often co-occurs, and also differs in being densely glandular-hairy and having gray-glaucous leaves; also co-occurs with *A. patula*, which has golden colored glands on short hairs, while *A. klamathensis* has dark colored glands on longer hairs.

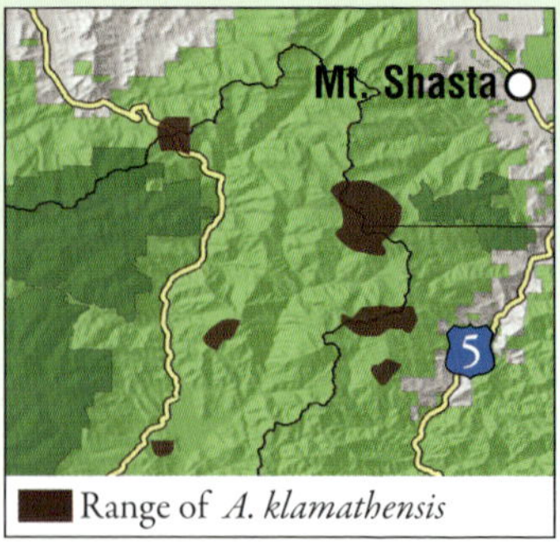

Santa Lucia Manzanita

Arctostaphylos luciana

FORM: shrub, small tree, 2-3 m STEM: twig sparsely short-nonglandular-hairy LEAVES: overlapped; blade 2-4 cm, 1.5-2.5 cm wide, ovate to ± round, glaucous, dull, appressed-gray-canescent, glabrous in age, base lobed, clasping, tip obtuse INFLORESCENCE: +- raceme, 0-1-branched; nascent inflorescence pendent, axis 0.5-1 cm, >1 mm wide, crowded, sparsely short-nonglandular-hairy; bracts 5-10 mm, leaf-like, lance-linear, canescent; pedicel 5--10 mm, glabrous FRUIT: 6-12 mm wide, depressed-spheric, glabrous; stones free HABITAT: shale outcrops, slopes, upland chaparral near coast; 400-800 m; East Cuesta Ridge Road from Cuesta Pass, east of Highway 101, San Luis Obispo County REMARKS local endemic with limited distribution, although within sight of the Pacific Ocean, it generally grows at the upper boundary of the marine layer in a warmer, drier climate than other auriculate-leaved species of the San Luis Obispo area.

Mallory's Manzanita

Arctostaphylos malloryi

FORM: shrub, erect, 1-3 m STEM: twig (and inflorescence bract) sparsely short-sticky- glandular-hairy LEAVES: erect; blade 2-3 cm, 1.5-2.5 cm wide, round to ovate, glaucous, dull, canescent or densely white-tomentose, glabrous in age, base rounded, truncate, or ± lobed, tip abruptly soft-pointed INFLORESCENCE: panicle, 2-5-branched; nascent inflorescence pendent, axis 1-2 cm, >1 mm wide; bracts 3-5 mm, scale-like, appressed; pedicel 6--9 mm, glandular-hairy FRUIT: 7-9 mm wide, depressed-spheric, glabrous to nonglandular-hairy; stones free HABITAT: volcanic soils, interior chaparral; 650-1200 m; South Fork and Iron Mountain areas of Shasta County; and Deer Lick Knob Rd. area north of Highwy 36 in Trinity County REMARKS: similar to *A. canescens* in the fine, dense-white hairs on its leaves and branchlets, differs in having short, glandular hairs and the short scale-like bracts on its panicles compared to the much longer, leafy bracts of *A. canescens*; fruit pedicels are consistently straight and not re-curved as is usually the case with *A. canescens*; may have arisen as a cross between *A. canescens* and *A. viscida* subsp. *pulchella*.

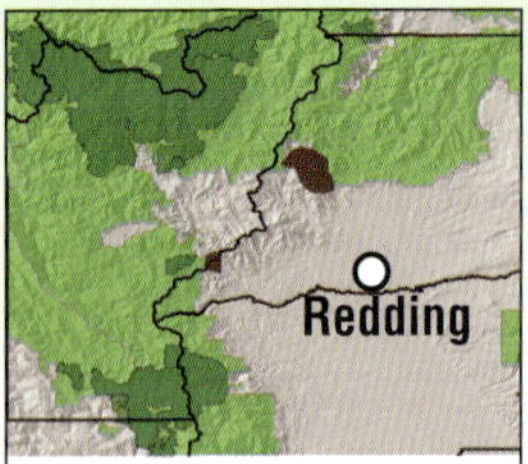

Range of *A. malloryi*

Common Manzanita

Arctostaphylos manzanita subsp. *manzanita*

FORM: erect, 2-8 m STEM: twigs smooth to finely tomentose LEAVES: erect; blade 2.5-5 cm, 1-3.5 cm wide, wide- or oblong-ovate dull to somewhat green, base rounded to ±wedge-shaped INFLORESCENCE: panicle, 2-7-branched; nascent inflorescence pendent, axis 1.5-4.5 cm; bracts appressed, 2-4 mm, scale-like, deltate to sharp-pointed; pedicel 3-8 mm, glabrous FRUIT: 8-12 mm wide, depressed-spheric, not glandular HABITAT: woodland, chaparral, conifer forest; 30–1200 m; widespread in inner North Coast Range and Sierra Nevada foothills REMARKS: A large, spreading manzanita of woodland habitat rather than chaparral.

Konocti Manzanita

A. manzanita subsp. *elegans*

INFLORESCENCE: ovary glandular hairy FRUIT: glandular hairy HABITAT: woodland, chaparral, conifer woodland, generally volcanic soils; 220-1850 m; n. Napa and s. Lake co. REMARKS: No burl, sticky, glandular fruit giving them a stippled appearance; subsp. *roofii* also occasionally has glandular fruit but main stems have burl; associates with *A. manzanita* subsp. *manzanita, A. stanfordiana* subsp. *stanfordiana, A. canescens* subsp. *canescens, A. glandulosa* subsp. *glandulosa,* and *A. viscida* subsp. *pulchella.*

Glaucous Manzanita

A. manzanita subsp. *glaucescens*

STEM: glabrous to sparsely glandular-hairy LEAF: glaucous, scabrous to glabrous or sparsely glandular-hairy INFLORESCENCE: 5-7 branched, sparsely glandular-hairy, ovary minutely glandular-hairy FRUIT: to glabrous or sparsely glandular-hairy HABITAT: chaparral, conifer forest; 150-600 m; interior outer Coast Range REMARKS: A variable subspecies with some similarity to *A. viscida.*

Contra Costa Manzanita

Arctostaphylos manzanita subsp. *laevigata*

LEAF: bright green INFLORESCENCE: 2-4 branched, branches generally 15 mm long HABITAT: chaparral, rocky outcrops; 240-1100 m; north Contra Costa Co. on hard rocky ridges on the western slope of Mt. Diablo REMARKS: large, erect manzanita with bright-green, shiny leaves, somewhat similar to northern forms of *A. pungens*.

Roof's Manzanita

A. manzanita subsp. *roofii*

FORM: erect shrub, prominent burl INFLORESCENCE: ovary glandular-hairy FRUIT: glandular- hairy HABITAT: chaparral, inner North Coast Range, 100-1000 m; also foothills of Butte County in Northern Sierra REMARKS: Only burl-forming subspecies of *A. manzanita*; distinguished from *A. patula* by its lack of glands on branchlets and inflorescence where it has fine, short, non-glandular hairs; can be confused with *A. patula subsp. gankini*; however, occurs lower in elevation.

Wieslander's Manzanita

A. manzanita subsp. *wieslanderi*

STEM: short-glandular hairy LEAVES: scabrous INFLORESCENCE: short-glandular hairy, ovary glabrous to hairy; FRUIT: glabrous to glandular-hairy HABITAT: chaparral, conifer forest; 300-1500 m; Shasta and Tehama Co. REMARKS: leaves scabrous, or rough to the touch, and mostly glandular hairy; generally grows at elevations between subsp. *manzanita* and the glandular *A. patula*.

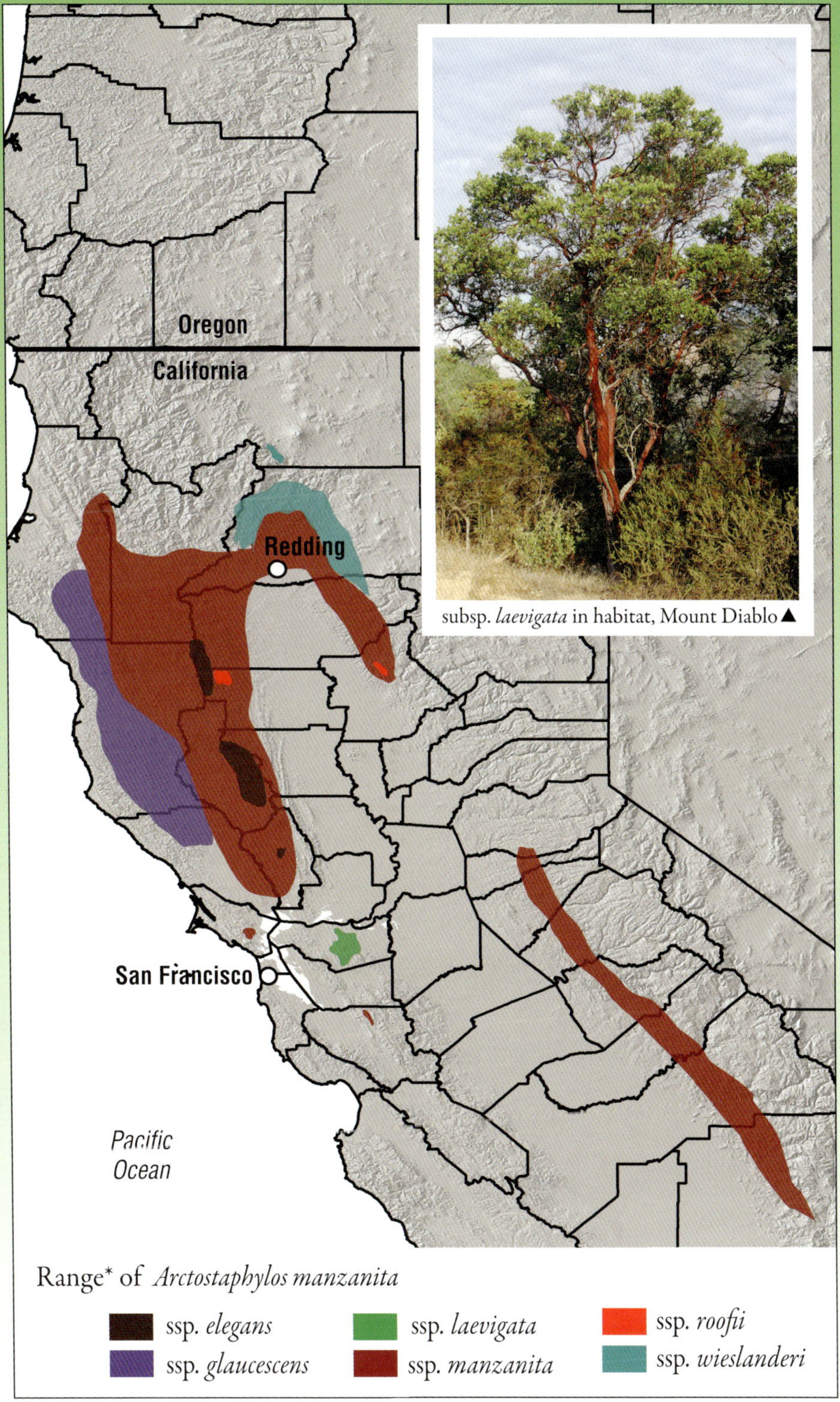

subsp. *laevigata* in habitat, Mount Diablo ▲

Range* of *Arctostaphylos manzanita*

Indian Manzanita

Arctostaphylos mewukka subsp. *mewukka*

FORM: shrub, erect, 1-3 m; burl STEM: twig glabrous LEAVES: blade 2-4 cm long, oblong- to lance-ovate, white-glaucous, tip acute INFLORESCENCE: panicle 2-7-branched; nascent inflorescence pendent, axis 2-3 cm, >1 mm wide, glabrous to +- nonglandular-hairy, occasionally glandular-hairy; bracts 3-6 mm, scale-like, deltate to lance-linear, glabrous; pedicel 4-8 mm, glabrous FRUIT: 10-16 mm wide, depressed-spheric, glabrous, dark mahogany-brown; stones generally free HABITAT: chaparral, forest openings; 333-1800 m; Central to Northern Sierra, North Coast Range, southern Cascades, (Moody Ridge, Placer County and Peavine Ridge, El Dorado County) REMARKS: often overlooked because of its superficial resemblance to *A. viscida* but its leaves are larger, even more glaucous-white, and are oval shaped compared to the rounded leaves of *A. viscida*; only white-glaucous leaved manzanita in the Sierra-Cascade Range that has a basal burl (*A. patula* often has a burl but it has bright green leaves).

Range of *A. mewukka* ssp. *mewukka*

True's Manzanita

Arctostaphylos mewukka subsp. *truei*

FORM: shrub erect, 2-4 m, no burl STEM: twig glabrous to ±nonglandular-hairy, rarely glandular pubescent LEAVES: blade 4-7 cm long, round to round-ovate, white-glaucous, tip obtuse INFLORESCENCE: panicle 4-7 branched, bracts appressed, hiding buds, keeled, tip acuminate FRUIT: stones free, occasionally some fused HABITAT: chaparral, forest openings; 290–1350 m; Paradise area of Butte County and Feather River Canyon REMARKS: differs from subsp. *mewukka* in absence of a basal burl, also growth form is tall, erect, and usually single-stemmed compared to the multi-stemmed, spreading, somewhat mound forming subsp. *mewukka*, leaf color is variable, from dull glaucous-green to gray-glaucous, although usually not as white-glaucous as subsp. *mewukka*; has perhaps the largest leaves of any manzanita species

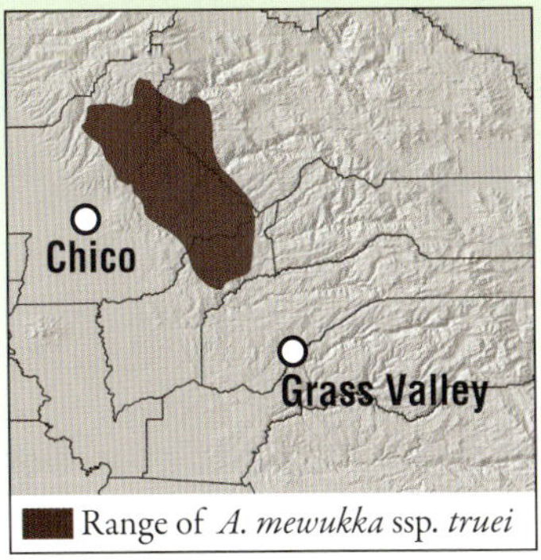

Mount Tamalpais Manzanita

Arctostaphylos montana subsp. *montana*

FORM: mounded to erect, 0.5-2 m STEM: twig densely short-white-nonglandular-hairy LEAVES: 1-1.5 cm wide, round-elliptic to elliptic; glabrous in age, dark green, shiny, base rounded to wedge-shaped, tip acute INFLORESCENCE: panicle, 1-3-branched; nascent inflorescence pendent, axis 0.5-1.5 cm, >1 mm wide, densely short-white-nonglandular-hairy; bracts appressed, scale-like, ovate, acuminate, 2-5 mm, glabrous; pedicel 3-6 mm, glabrous FRUIT: 6-8 mm wide, spheric, glabrous; stones free HABITAT: narrowly endemic species found in serpentine chaparral and Sargent cypress chaparral woodland; Mount Tamalpais and ridges to the north in Marin County REMARKS: with dense white non-glandular hairs on its branchlets and inflorescence, similar to *A. canescens* (with which it associates), also with short, brown-colored scaly bracts that do not recurve like *A. hookeri* and *A. pungens*; more similar to prostrate subsp. *ravenii* (which has round, shiny leaves), but subsp. *montana* is larger, mounding to erect, and with more elliptic leaves.

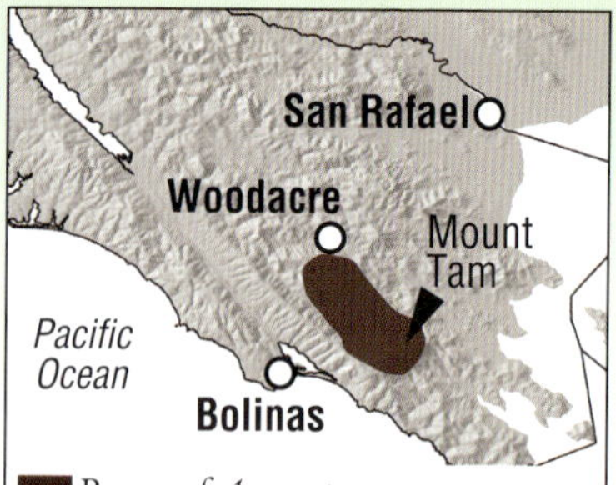

Range of *A. montana* ssp. *montana*

Presidio Manzanita

Arctostaphylos montana subsp. *ravenii*

FORM: prostrate, 0.1-0.3 m STEM: twig short- gray-nonglandular-hairy LEAVES: blade 1-2 cm, round to round-elliptic, dark green, shiny FRUIT: 4-5 mm wide HABITAT: serpentine coastal bluff and prairie; 60-95 m; several of the Bay Area botanical gardens, including Tilden, have it cultivated REMARKS: differs from *A. franciscana* in its prostrate nature and thick, round, shiny-green leaves; most rare of all manzanitas with all living plants today (both in cultivation and in the wild) originating from a single individual found on the Presidio by Peter Raven in 1951. Both subsp. *ravenii* and subsp. *montana* are tetraploids, unlike the closely related *A. franciscana* which is diploid, and this is partly why they have been grouped together as subspecies.

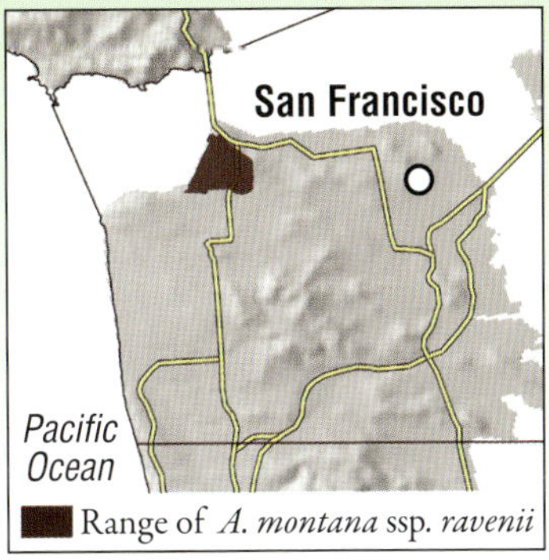

Montara Mountain Manzanita

Arctostaphylos montaraensis

FORM: mounded to erect, 0.5-5 m (<0.5 m on exposed granite outcrops) STEM: twig (and immature inflorescence axis) densely long-glandular-hairy LEAVES: overlapped; blade 2.5-4.5 cm, 1.5-2.5 cm wide, ovate, bright green, sparsely glandular-puberulent, base lobed, clasping, tip acute INFLORESCENCE: panicle, 4-6-branched; nascent inflorescence pendent, axis 1-1.5 cm, >1 mm wide; bracts 6-9 mm, leaf-like, lanceolate, acuminate, glandular-hairy; pedicel 5-6 mm, glandular-hairy FRUIT: 6-7 mm wide, depressed-spheric, glandular-hairy, sticky; stones free HABITAT: granite, sandstone outcrops, chaparral, coastal scrub; 200-500 m; northern San Mateo County including Sand Bruno Mountain and Montara Mountain. REMARKS: tall, erect shrub, can be relatively prostrate on rocky granite ridges (where it resembles *A. imbricata*); *A. montaraensis* associates with *A. crustacea* subsp. *crustacea*; one of seven San Francisco Bay area manzanitas characterized by leafy bracts and basal-lobed auriculate-leaves.

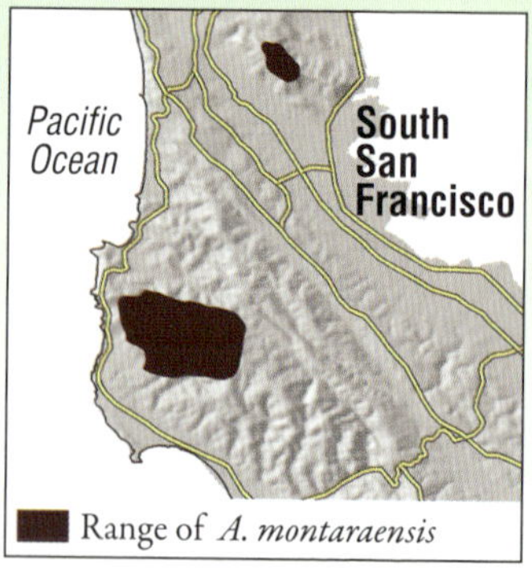

Monterey Manzanita

Arctostaphylos montereyensis

FORM: erect, 1-3 m STEM: twig (and leaves, nascent inflorescence axis, bractS, pedicel, ovary, fruit) glandular-hairy LEAVES: erect, not overlapped; blade 2-3 cm, 1.5-3 cm wide, round to oblong-ovate, bright green or ±glaucous, dull, glandular-hairy, papillate, scabrous, base generally rounded, tip abruptly soft-pointed INFLORESCENCE: panicle, 4-10-branched; nascent inflorescence pendent, axis 1-2 cm, > 1 mm wide; bracts 3-12 mm, leaf-like, green, persistent, lanceolate, acuminate; pedicel 5-6 mm FRUIT: 8–12mm wide, depressed-spheric, sticky; stones free HABITAT: sandstone soils (stabilized dunes), chaparral <350 m; eastern Fort Ord and on slopes of nearby Toro Mountain REMARKS: dull green glandular-hairy leaves are scabrous (rough to the touch) and maroon-tinged leafy bracted inflorescences are densely covered in glandular hairs; restricted endemic associates with *A. tomentosa*, *A. pumila*, *A. hookeri*, and rarely *A. pajaroensis*; one of four manzanita endemics found exclusively on foothills surrounding Monterey Bay; with its large, twisted trunks it is an especially attractive manzanita.

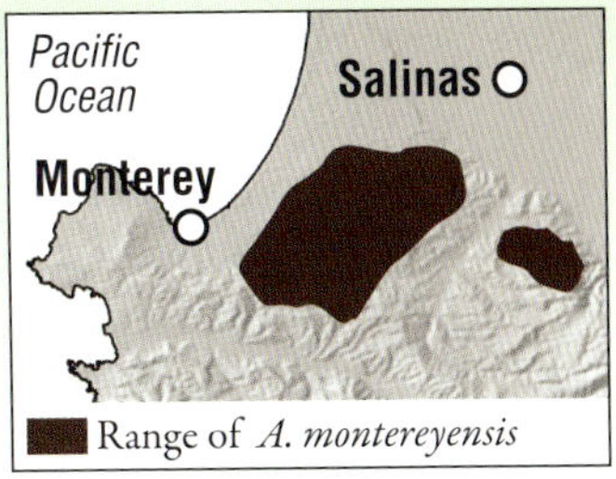

Moran's Manzanita

Arctostaphylos moranii

FORM: shrub, 1-2.5m tall with basal burl STEM: twigs with short, appressed hairs LEAVES: on petioles 5-10 mm long; narrowly elliptic or lance-elliptic to ovate; gray-glaucous from waxy deposit INFLORESCENCE: panicle 4-6 branched, lax, elongated horizontally, spreading with flowers well-spaced; bract leaf-like but reduced, acuminate, 4-7 mm long, pedicels and ovary glandular-hairy FRUIT: 4-6 mm wide, glabrescent; nutlets mostly coalesced as solid or in part separable HABITAT: coastal scrub, chaparral; above the village of San Antonio de las Minas northeast of Ensenada REMARKS: shares common characters with *A. australis*, such as the thin, erect, openly spreading immature panicles with buds widely spaced and fruit with nutlets typically fused into a solid stone; differs by having a burl and being densely glandular-hairy; originally described as a variety of *A. australis*; resembles *A. australis* and *A. glandulosa* in morphologically.

Range of *A. moranii*

Morro Manzanita

Arctostaphylos morroensis

FORM: erect, 1-4 m STEM: old stems gray, bark shredding; twig short-and long-white-nonglandular-hairy LEAVES: overlapped; blade 1.5-3 cm, 1-2cm wide, oblong-ovate to -elliptic; dark green, shiny, and lacking stomata above, densely tomentose below; base ± truncate to ±cordate, tip abruptly soft-pointed INFLORESCENCE: panicle, 2-5-branched; nascent inflorescence pendent, bell-shaped, axis 0.5-0.8 cm, > 1 mm wide; bracts 5-8 mm, leaf-like, lance-linear, acuminate, puberulent; pedicel 4-6 mm, glabrous or nonglandular-hairy FRUIT: 7–10 mm wide, depressed-spheric; stones free HABITAT: stabilized sand dunes, sandstones, chaparral; <200 m; Morro Bay region; Montana de Oro State Park REMARKS: large, spreading manzanita with twisted trunks of gray shreddy bark instead of the typical smooth red bark (bark similar to *A. tomentosa* along with its bifacial leaves); leaf base is truncate and sometimes cordate, but it does not have the auriculate-clasping leaf bases of other nearby species such as *A. pechoensis, A. osoensis,* and *A. cruzensis*; with its backdrop of Hollister Peak to the east and Morro Rock to the north, *A. morroensis* grows in especially scenic locations.

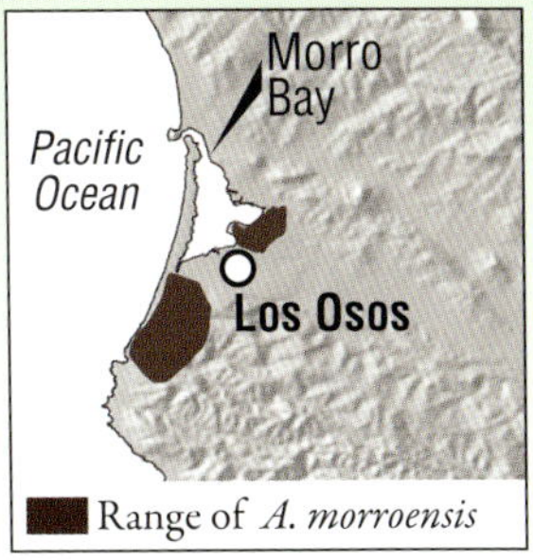

Ione Manzanita

Arctostaphylos myrtifolia

FORM: mounded to erect, 0.5-1.5 m STEM: old stems smooth, red with ±gray or ± glaucous patches; twig glandular-hairy LEAVES: spreading; blade 0.6-1.5 cm, 0.3-0.8cm wide, narrowly elliptic, bright green, shiny, sparsely glandular-hairy, papillate, base wedge-shaped, flat INFLORESCENCE: +- raceme, 0-1-branched; nascent inflorescence pendent, axis 0.5-1 cm, >1 mm wide, glandular-hairy; bracts 1-2 mm, scale-like, deltate, glabrous; pedicel 1-3 mm, glabrous FRUIT: small, 3-4 mm wide, ±spheric, glabrous; stones free HABITAT: acidic sandy or clay soils, chaparral, woodland; 70-770 m; Amador and Calaveras counties near Ione in acidic sandy soil REMARKS: slender panicles with reduced bracts and small, pulpless fruit that falls very early while still apparently developing, and therefore the fruit are rarely observed; shiny green leaves are among the smallest in the genus; smooth red, or often uniquely gray-glaucous colored bark; branchlets are covered with long, glandular hairs; once grouped with *A. sensitiva* because of its early fruit drop but molecular genetic studies suggest that it is not closely related to any particular species; one of only two localized endemic species in the Sierra Nevada (the other is *A. nissenana*).

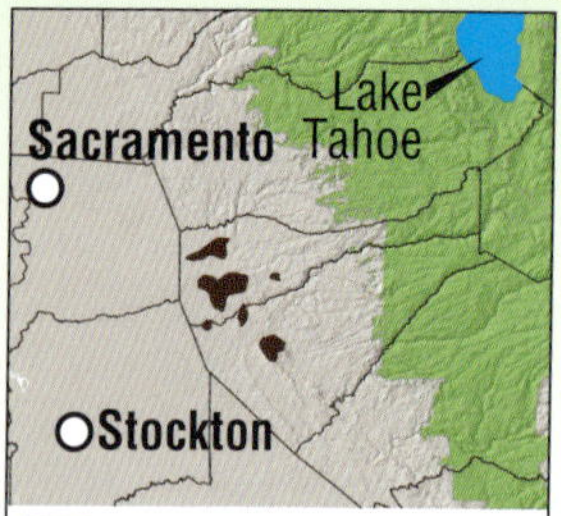

Range of *A. myrtifolia*

Pinemat Manzanita

Arctostaphylos nevadensis subsp. *nevadensis*

FORM: prostrate to mounded, 0.1-0.5 m; burl absent STEM twig (and nascent inflorescence axis) sparsely short-nonglandular-hairy LEAVES: erect; blade 1-3 cm, 1-1.5 cm wide, obovate or oblanceolate, bright green, shiny, puberulent, glabrous in age, base wedge-shaped, tip obtuse or acute INFLORESCENCE: +- raceme, 0-1-branched; nascent inflorescence pendent, spheric, axis 0.5-1 cm, >1 mm wide; bracts 2-3 (5-10) mm, scale-like, linear or lance-linear, acuminate, glabrous to sparsely nonglandular-hairy (rarely glandular-hairy); pedicel 3-5 mm, glabrous FRUIT: 6-8 mm wide, spheric to depressed-globose, dark red; stones generally free HABITAT: rocky outcrops, conifer forest; 1,300–3,200 m (usually above 2,500 m in the south, and 1,300 m in the north); higher elevation conifer forests across California, Oregon, and Washington REMARKS: hybrids between *A. nevadensis* and *A. patula* occur sporadically; in Washington, it is largely restricted to serpentine soils, similar to subsp. *knightii* in the Klamath Region of California; similar to *A.* X *coloradensis*, likely a hybrid between *A. patula* and *A. uva-ursi*.

Knight's Pinemat Manzanita

Arctostaphylos nevadensis subsp. *knightii*

FORM: prostrate to mounded, 0.1-0.5 m; burl present, occasionally obscure HABITAT: rocky outcrops, serpentine, conifer forests; 100-2600 m; vicinity of High Divide Road, Del Norte Co. and Bald Hills area of Humboldt County, California REMARKS: associates with *A. hispidula* and *A. nortensis*; the primary difference between *A. nevadensis* subsp. *knightii* and subsp. *nevadensis* is that subsp. *knightii* has small rounded burls and resprouts after a fire; appears to extend up into Southwestern Oregon.

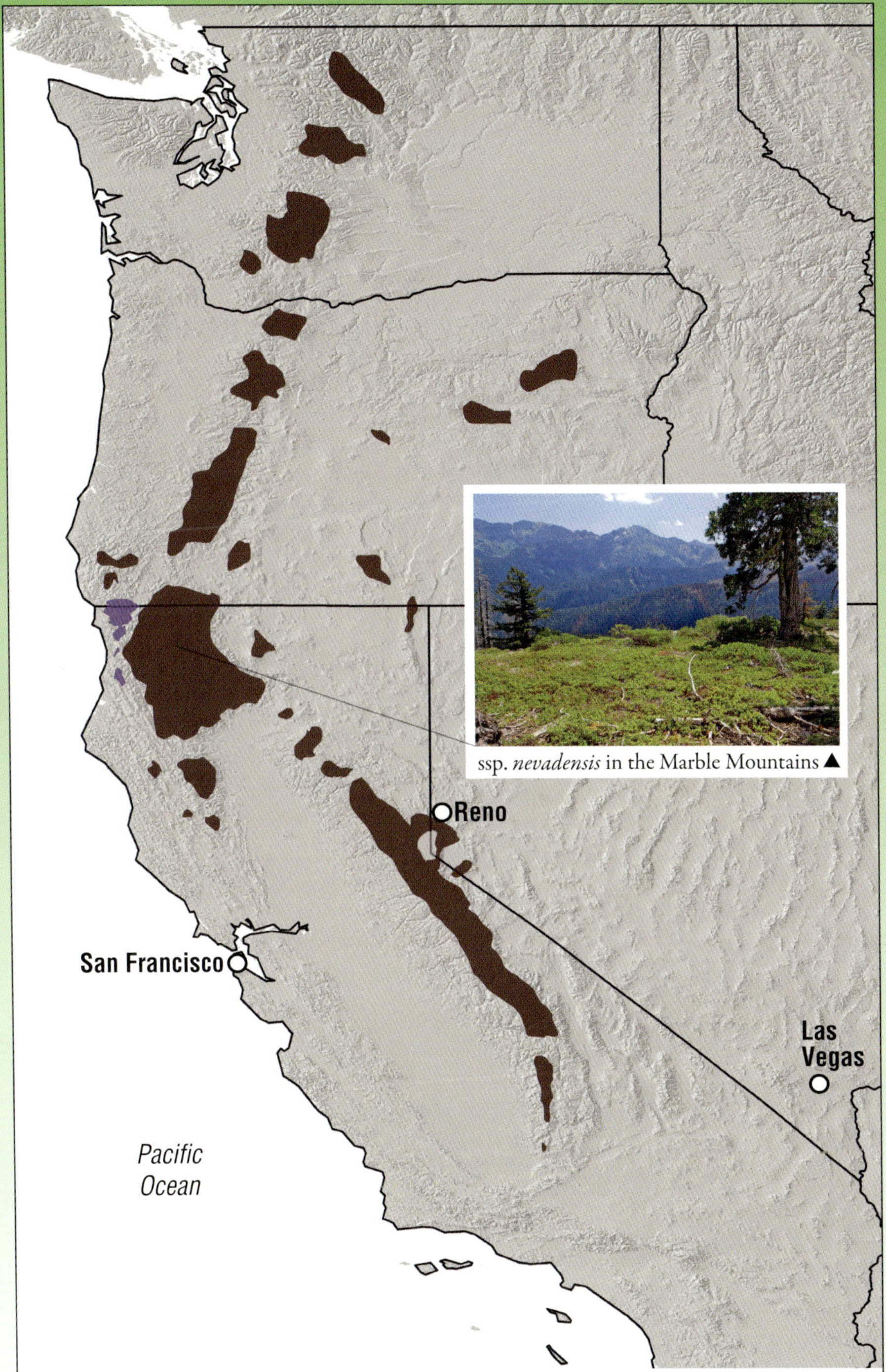

Range of *Arctostaphylos nevadensis* — subsp. *nevadensis*; subsp. *knightii*

El Dorado Manzanita

Arctostaphylos nissenana

FORM: erect, 0.2-1.5 m STEM: old stem bark persistent, gray, shredding; twig canescent LEAVES: erect; blade 1-2 cm, 0.8-1.5 cm wide, elliptic to oblong-elliptic, glaucous, gray, dull sparsely appressed-puberulent, glabrous in age, base wedge-shaped to rounded, tip acute INFLORESCENCE: +- raceme, 0-1-branched; nascent inflorescence pendent, axis 0.2-0.5 cm, >1 mm wide, canescent; bracts 3-5 mm, leaf-like, lanceolate, glabrous to sparsely nonglandular-hairy; pedicel 5-7 mm, glabrous to sparsely nonglandular-hairy FRUIT: 3-4 mm wide, ±cylindric, glabrous, mature splitting, stones free HABITAT: open rocky shale ridges, chaparral, woodland, 450-1650 m; for example, vicinity of Poho Ridge in El Dorado County near Pollock Pines with *A. viscida* and *A. mewukka* REMARKS: only species in Sierra Nevada with gray-shreddy bark instead of smooth red bark; distinguished from *A. viscida* by its shaggy bark and smaller leaves, which are canescent and a dusty blue-green compared to glabrous white-glaucous of *A. viscida*, also blooms a month earlier than *A. viscida*; small tan fruits fall quickly and are usually gone by mid-summer while those of *A. viscida* can be found in early winter.

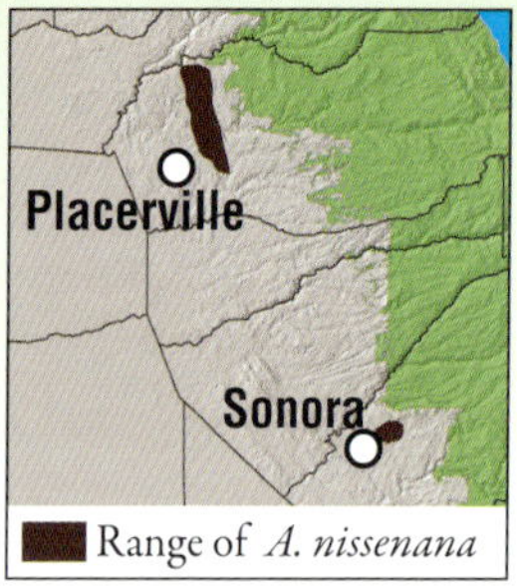

Del Norte Manzanita

Arctostaphylos nortensis

FORM: erect, 2-5 m STEM: twig (and inflorescence bract) long-nonglandular-bristly, occasionally minutely glandular-hairy LEAVES: erect; blade 2-4 cm, 1-2 cm wide, narrowly elliptic-ovate, glaucous, gray, dull, white-canescent, in age dark green, dull, glabrous, base truncate to rounded, tip acute INFLORESCENCE: +- raceme, 0-1-branched; nascent inflorescence pendent, axis 0.5-1.2 cm, >1 mm wide, soft-nonglandular-hairy and long-nonglandular-bristly, occasionally minutely glandular-hairy; bracts 5-10 mm, leaf-like, ovate, acute, green, persistent; pedicel 2-4 mm, sparsely white-nonglandular-hairy FRUIT: 6-8 mm wide, depressed-spheric; stones free HABITAT: rocky slopes, occasionally serpentine, chaparral, conifer forest; 500-940 m; north Klamath range, for example, near Gasquet, California REMARKS: differs from *A. columbiana* in being less-tree like with smaller, more elliptic shaped leaves that are glaucous-gray and canescent; branchlets can have glandular hairs or non-glandular white hairs, also combining characters of both *A. columbiana* and *A. canescens*; associates with *A. hispidula* and *A. nevadensis* subsp. *knightii*; growing in an area between the coastal *A. columbiana* and west of the more inland *A. canescens*, it has been speculated that this species arose from a cross between these two species.

Glossyleaf Manzanita

A. nummularia subsp. *nummularia*

FORM: prostrate to erect, 0.5-5 m STEM: old stem bark gray, shredding LEAVES: blade 1-2.2 cm, 0.8-1.8 cm wide, round to round-ovate, base truncate to ±lobed; dark shiny green on top and light green on bottom; no stomata on upper leaf surface INFLORESCENCE: panicle, 4-12-branched; nascent inflorescence pendent, axis 0.5-1 cm, <1 mm wide; bracts appressed, 0.5-2 mm, scale-like, deltate, sharp-pointed, glabrous; pedicel 3-5 mm, glabrous FLOWERS: flowers have four sepals and four lobes on the corollas while other manzanitas have sepals and corolla lobes in fives FRUIT: 3-4 mm wide, ± cylindric, glabrous, mature splitting; stones free HABITAT: coastal prairie, maritime chaparral, closed-cone forest; 15-400 m: for example, Salt Point State Park, coastal areas from extreme northern Sonoma County to central Mendocino County REMARKS: branchlets and inflorescence have dense, short non-glandular hairs, along with longer gland-tipped hairs; fruit are very small and pulpless and the exocarp (skin) becomes wrinkled before falling, fruit are rarely seen as they fall very quickly in early summer before they appear to be fully developed; except for for bark character and leaf size, similar to *A. sensitiva* to the south.

Pygmy Manzanita

A. nummularia subsp. *mendocinoensis*

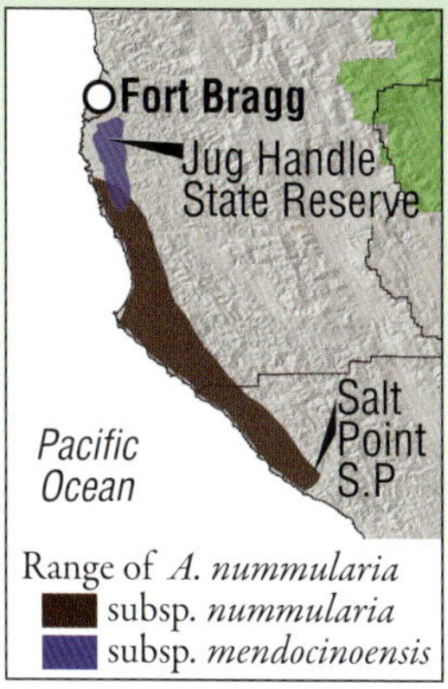

FORM: prostrate to mounded, 0.1-0.5 m STEM: old stem bark red-gray, generally rough or shredding LEAVES: blade 0.5-1.2 cm, 0.3-0.7 cm wide, oblong-elliptic, base rounded to wedge-shaped ±cylindric, glabrous, mature splitting HABITAT: pygmy pine forest, chaparral; 50-200 m; Jug Handle State Reserve, found primarily in the Fort Bragg-Mendocino area REMARKS: leaves are the smallest in the genus and comparable only in size to *A. myrtifolia* (slightly longer leaves); four sepals and petals per flower, and the small, pulpless fruit fall very early and are rarely seen; this subspecies integrates into taller forms of *A. nummularia* subsp. *nummularia* outside acid-clay soils of the pygmy forests near Fort Bragg.

Bishop Manzanita

Arctostaphylos obispoensis

North

South

FORM: erect, 1-4 m STEM: twig (and nascent inflorescence axis) sparsely short-nonglandular-hairy LEAVES: erect; blade 2-4.5 cm, 1-2.5 cm wide, oblong- to lance- ovate, glaucous-gray, dull, appressed-canescent, in age glabrous, base rounded to truncate or ±lobed, tip acute INFLORESCENCE: panicle, 2-4-branched; nascent inflorescence pendent, bell-shaped, axis 1-2.5 cm, >1 mm wide; bracts 7-14 mm, leaf-like, lance-linear, acuminate, appressed-canescent to glabrous; pedicel 8-10 mm, glabrous FRUIT: 9-14 mm wide, depressed-spheric, glabrous; stones free HABITAT: rocky, generally serpentine soils, chaparral, open closed-cone forest near coast in central-southern Santa Lucia Mountains; 60-950 m; for example, W. Cuesta Ridge REMARKS: beautiful, gray colored foliage has hairs that tend to wear off with age becoming smooth; in the northern part of its range, species has leaves with rounded bases and in the southern end of its range plants with auriculate leaves are common and are similar to the nearby *A. luciana* (with leaves that are always strongly auriculate); La Cuesta Ridge is a place where range nearly overlaps with *A. pilosula* (east side) and *A. luciana* (south).

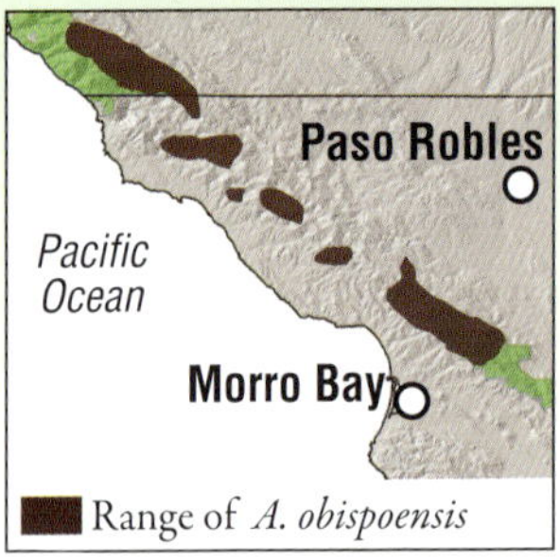

Ohlone Manzanita

Arctostaphylos ohloneana

FORM: erect, 1-4 m STEM: twig (and immature inflorescence axis) sparsely short-nonglandular-hairy, sometimes long glandular-hairy LEAVES: erect; blade 1.5-3 cm, 1-1.5 cm wide, elliptic to ovate-elliptic, puberulent and short-glandular-hairy, in age glabrous, light green, dull, base wedge-shaped INFLORESCENCE: panicle, 3-5-branched; nascent inflorescence pendent, axis 1.5-2.5 cm, >1 mm wide; bracts 1-3 mm, scale-like, ovate to deltate, sharp-pointed, nonglandular-hairy; pedicel 3-8 mm, glabrous FRUIT: 5-8 mm wide, depressed-spheric, glabrous; stones free HABITAT: siliceous shale outcrops, chaparral, knobcone-pine woodland, 400-500 m; only on private property in the Santa Cruz Mountains (Lockheed Chalks); REMARKS: distinctive, recurved floral bracts on its immature inflorescence; different from any other manzanita species found in the Santa Cruz Mountains; four small populations are embedded within the larger population of *A. glutinosa*, another manzanita restricted to the Lockheed Chalks; also associates with *A. andersonii, A. crustacea* subsp. *crinita, A. silvicola*; most recently discovered manzanita species, described in 2008; diploid, small clade, likely to be of relatively ancient origin.

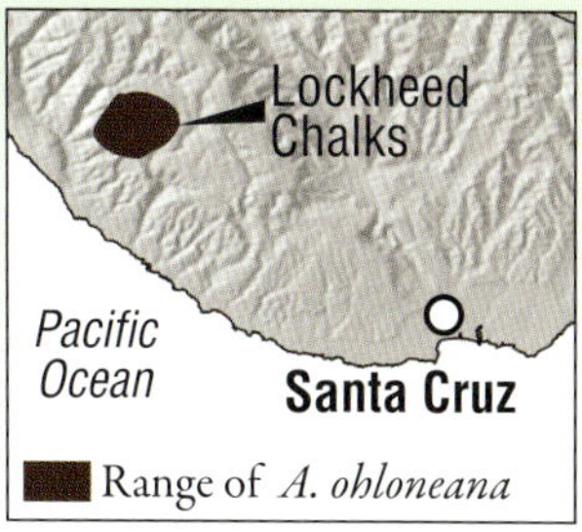

Oso Manzanita

Arctostaphylos osoensis

FORM: erect, 1-4 m STEM: bark persistent, gray, shredding; twig (and nascent inflorescence axis) sparsely short-nonglandular-hairy LEAVES: overlapped; blade 1.5-3 cm, 1.5-2.5 cm wide, ovate to round-ovate, dark green, ±shiny, sparsely short- nonglandular-hairy or glabrous, base deeply lobed, clasping, tip obtuse INFLORESCENCE: +- raceme, 0-1-branched; nascent inflorescence pendent, short-nonglandular-hairy, axis 0.5-1 cm, > 1 mm wide; bracts 4-8 mm, leaf-like, lanceolate to ovate, glabrous; pedicel 8-9 mm, glabrous FRUIT: 5-8 mm wide, depressed-spheric, glabrous; stones free HABITAT: dacite (volcanic) outcrops, chaparral; 50–375 m OBSERVATIONS: Hollister Peak, and on adjacent ridges, ~13km west of San Luis Obispo near Los Osos State Park REMARKS: differs from *A. pechoensis*, another auriculate-leaved species growing nearby, in that it lacks the long, white bristly hairs of that species and presents only short, nonglandular hairs; also with smaller fruit (5-8mm) with separable stones compared to the larger (8-12mm), spherical fruit with fused stones of *A. pechoensis*; auriculate leaves clasp the branchlets and have bright red edges on new growth; variable with some specimens lacking shreddy bark; mountains surrounding San Luis Obispo are a hotspot for manzanita diversity.

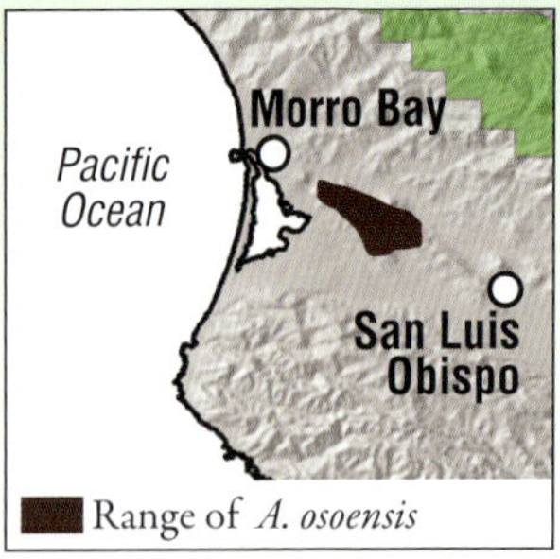

Otay Manzanita

Arctostaphylos otayensis

FORM: erect, 1-3 m STEM: twig sparsely short- nonglandular- and long-glandular-hairy LEAVES: erect; blade 2-3 cm, 1-2 cm wide, narrowly or oblong-elliptic to ovate, gray-glaucous, dull, appressed-tomentose, in age ± glabrous, base wedge-shaped, rounded, or truncate, tip abruptly soft-pointed INFLORESCENCE: panicle, 4-7-branched; nascent inflorescence pendent, axes 1.5-2.5 cm, > 1 mm wide; bracts spreading, 6-12 mm, leaf-like, rigid, lance-linear, acuminate, green, long-glandular-hairy, persistent; pedicel 3-8 mm, glandular-hairy FRUIT: 6-8 mm wide, spheric, sparsely nonglandular-hairy to glabrous; stones free or fused HABITAT: volcanic rock outcrops, chaparral, woodland; 280-1700 m; south Peninsular Range, e.g., Otay Mountain and both Jamul and San Miguel mountains just to the north REMARKS: similar to *A. bolensis* in that both species are densely glandular-hairy with gray-glaucous leaves, have glabrous or smooth fruit; like many southern California and northern Baja species, the stones of *A. bolensis* are consistently fused while those of *A. otayensis* are partially to fully fused; may also exist on some of the nearby mountains across the border in Baja California, more exploration needed.

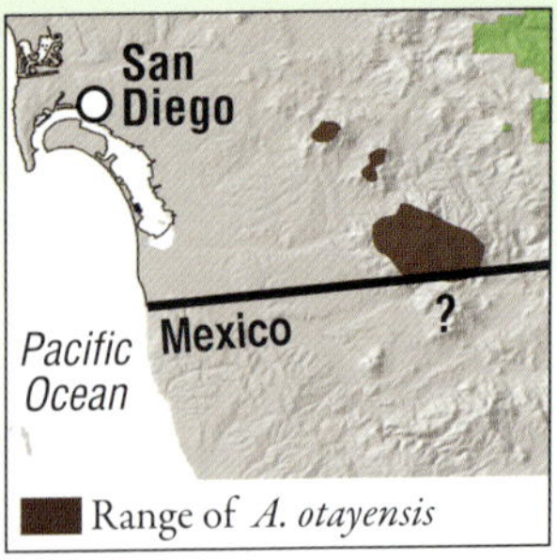

Pacific Manzanita

Arctostaphylos pacifica

FORM: prostrate, 0.1-0.6 m; obscure burl STEM: twig short-nonglandular-hairy LEAVES: spreading; blade 1-2 cm, 0.5-1 cm wide, elliptic; stomata on both surfaces with lower-leaf light green, upper-leaf dark green, sparsely nonglandular-hairy, base wedge-shaped, tip acute, margin serrate INFLORESCENCE: +- raceme, 0-1-branched; nascent inflorescence pendent, short-nonglandular-hairy, axis 0.5-1 cm, >1 mm wide; bracts 0.5-1 mm, scale-like, awl-shaped, glabrous; pedicel 3-5 mm, glabrous FRUIT: 6-8 mm wide, spheric, ± red, glabrous; stones free HABITAT: sandstone outcrops, chaparral; 300 m; Pacific Rock, San Bruno Mountain County Park; REMARKS: two distinct, small prostrate patches associated with *A. uva-ursi, A. imbricata, A. montaraensis*, and a suspected hybrids between *A. uva-ursi* and *A. montaraensis*; many puzzling features, such as stomata on both surfaces unlike *A. uva-ursi* and consistently serrated leaves that are darker green on the upper surface and lighter green on the lower surface; with only one small location, consisting of two different plants, *A. pacifica* is one of the rarest of all manzanitas; needs further study.

Pajaro Manzanita

Arctostaphylos pajaroensis

FORM: erect, 1-4 m STEM: old stems generally with gray bark, shredding; twigs short-nonglandular-hairy (hairs occasionally long) LEAVES: overlapped; blade 2-4 cm, 1-2 cm wide, ovate to triangular-ovate; lower-leaf light green, glabrous, midvein nonglandular-hairy, upper-leaf lacking stomata, dark blue-green, glaucous; base lobed, clasping, tip acute, margin tinged with red INFLORESCENCE: panicle, 2-5-branched; nascent inflorescence pendent, short-nonglandular-hairy, axis 1-1.5 cm. >1 mm wide; bracts 5-10 mm, leaf-like, lance-linear, acuminate, glabrous; pedicel 5-8 mm, generally nonglandular-hairy FRUIT: 6-8 mm wide, depressed-spheric, sparsely nonglandular-hairy or glabrous; stones free HABITAT: sandstone outcrops, chaparral; <755 m; occurs primarily in Monterey County, for example, Manzanita Park near Prunedale with *A. hookeri* subsp. *hookeri* and *A. crustacea* subsp. *crustacea* REMARKS: only manzanita that has the combination of shreddy bark and auriculate leaves with different surfaces; only other species with similar leaves is *A. andersonii* which is distinguished by smooth, red bark; local endemic, impacted by suburban development; restricted to relatively few preserves growing in and around residential areas.

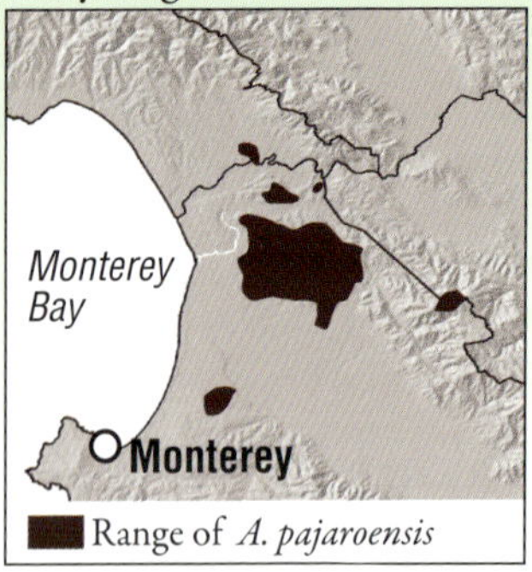

Alameda Manzanita

Arctostaphylos pallida

FORM: erect, 2-4 m STEM: twig densely short- and long-nonglandular-hairy LEAVES: overlapped; nonglandular-hairy; blade 2.5-4.5 cm, 2-3 cm wide, generally ovate, glaucous-green, dull, glabrous, base lobed, clasping, tip acute INFLORESCENCE: panicle, 3-5-branched; nascent inflorescence pendent, axis 0.5-1 cm, >1 mm wide, densely short-nonglandular-hairy, occasionally with long glandular hairs; bracts 5-9 mm, leaf-like, widely lanceolate, acute, glandular-hairy; pedicel 8-12 mm, glandular-hairy FRUIT: 8-10mm wide, depressed-spheric, sticky; stones free HABITAT: siliceous shales, slopes, ridges, chaparral; 200-460 m, East San Francisco Bay Hills, for example Huckleberry Botanic and Sobrante Ridge regional preserves (surrounded by development) REMARKS: "pallida" refers to the bloom on the young leaves, giving them a pale green color; an extremely short petiole along with closely imbricated, heart-shaped leaves, creates a tight clasp to the branchlets; associates with *A. crustacea* subsp. *crustacea* and coast live oak.

Range of *A. pallida*

Parry's Manzanita

A. parryana subsp. *parryana*

FORM: mounded,spreading, generally 0.5-1m, lacks a burl LEAVES: green, erect; blade 1.5-5 m, 1.5-2.5 cm wide, oblong-ovate to widely elliptic, glabrous or sparsely nonglandular-hairy, base obtuse to truncate, tip acute STEM: twig sparsely to moderately short-nonglandular-hairy FLOWERS: ovary glabrous INFLORESCENCE: panicle, 3-5-branched; nascent inflorescence pendent, moderately short-nonglandular-hairy 0.5-1.5 cm, >1 mm wide; bracts +- appressed, 2-4 mm, scale-like, deltate to wide ovate, glabrous; pedicel 4-9 mm, glabrous FRUIT: 7-10 mm wide, spheric, glabrous; stones fused HABITAT: openings in chaparral, pinyon-juniper forest; 1500-2400 m; western Transverse Range, for example Mount Pinos REMARKS: spreading or mounding habit (occasionally prostrate) with small leaves that are bright green, shiny, and usually smooth or with sparse, short hairs; sometimes associates with *A. patula*, which has narrower, more elliptic leaves with distinct, small, golden glands; originally described as a hybrid between *A. pungens* and *A. glauca*; one individual has been identified as *A. parryana* at the UC Granite Mountain Reserve in the eastern Mojave Desert.

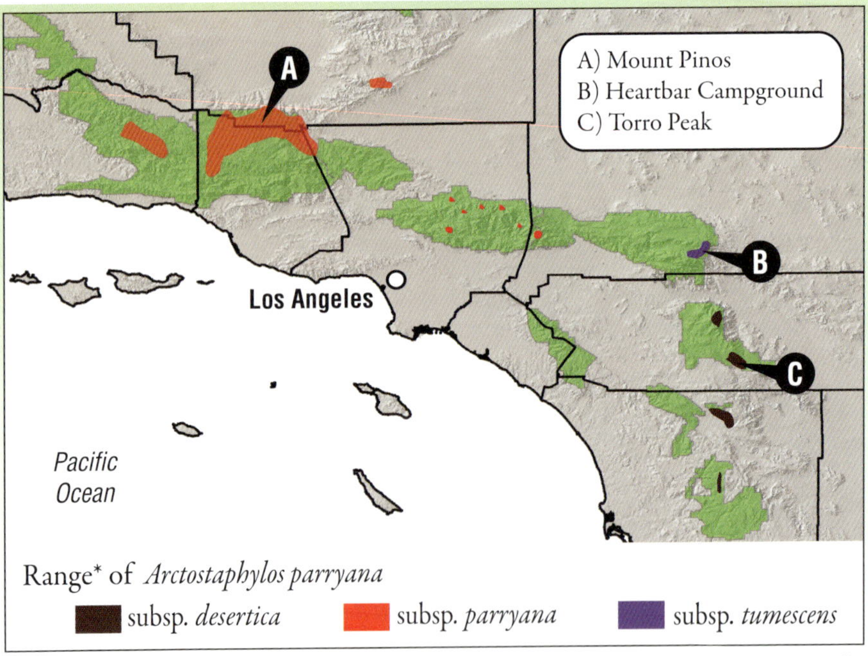

Desert Manzanita
A. parryana subsp. *desertica*

FORM: erect, 1-2 m; burl LEAVES: gray-glaucous, narrow and pointed HABITAT: chaparral; 1200-2300 m; Peninsular Ranges, for example, San Jacinto, Santa Rosa, San Ysidro mountains REMARKS: associates with *A. glandulosa* subsp. *adamsii* and is similar in having branchlets with sparse, non-glandular hairs; differs in fruit that are larger and spheric, not depressed-spheric, with a solid stone rather than separable nutlets; differs from *A. parryana* subsp. *tumescens* in having intensely glaucous leaves that are narrow and more pointed with slightly smaller fruit; described in 1997 at the same time as subsp. *tumescens*, both distinguished from subsp. *parryana* by having a well-developed basal burl; collections of specimens in fruit (not flowers) was key to recognizing difference from *A. glandulosa* subsp. *adamsii*.

Interior Manzanita
A. parryana subsp. *tumescens*

FORM: erect, 1-2 m; burl prominent LEAVES: green, erect; blade 1.5-5 m, 1.5-2.5 cm wide, oblong-ovate to widely elliptic, STEM: twig (and nascent inflorescence axis) sparsely to moderately short-nonglandular- hairy FRUIT: stones fused HABITAT: montane chaparral; 2100-2300 m; open slopes above the campground on Heartbar Peak in the San Gabriel Mountains REMARKS: distinguished from *A. patula* by its lack of glandular hairs, having only fine, white, non-glandular hair (both have burls); fruit of subsp. *tumescens* is larger and more spheric with a solid stone while those of *A. patula* have separable nutlets; described as a subspecies in 1997; needs further study.

Greenleaf Manzanita

A. patula subsp. *patula*

FORM: erect, 1-3 m, with or lacking obscure basal burl STEM: twig short-gold-glandular hairy LEAVES: erect; blade 2.5-6 cm, 1.5-4 cm wide, widely ovate to round, bright green, shiny, glabrous, base rounded, truncate, or ±lobed, tip abruptly sofi pointed INFLORESCENCE: panicle, 2-8-branched; nascent inflorescence pendent, short-gold-glandular hairy, axis 1.5-3 cm, >1 mm wide; densely short-glandular-hairy; bracts 4-6 mm, scale-like, deltate, acuminate, glabrous; pedicel 2-7 mm, glabrous FRUIT: 7-10 mm wide,±spheric or depressed-spheric, glabrous; stones free or partially fused HABITAT: montane chaparral, conifer forest; 750-3350 m; higher elevation conifer forests across Oregon, Washington, Montana, Colorado, New Mexico, Arizona, Utah, and Baja California REMARKS: bright-green shiny leaves and fine golden glands on branchlets and inflorescence; flowers are predominantly pink; one of the most widespread manzanitas in the western United States; sprouting habitat restricted to southern range (California) where it reaches its southern limit in the Sierra de San Pedro Mártir National Park in northern Baja.

Gankin's Manzanita

A. patula subsp. *gankinii*

FORM: burl hidden or often flat, obscure STEM: densely short-nonglandular-hairy INFLORESCENCE: characterized by silvery green hue from a distance whereas subsp. *patula* is more golden green; densely short-nonglandular-hairy HABITAT: montane chaparral, conifer forest; 750-3350 m; Sierra Nevada, Cascade, and Klamath Mountains of California and sw Oregon REMARKS: lacking fine golden glands on its branchlets and inflorescences; characterized by leaves that appear grey from a distance compared to the golden green hue of subsp. *patula*; described as a subspecies in 2019; can be confused with *A. manzanita* subsp. *roofii*.

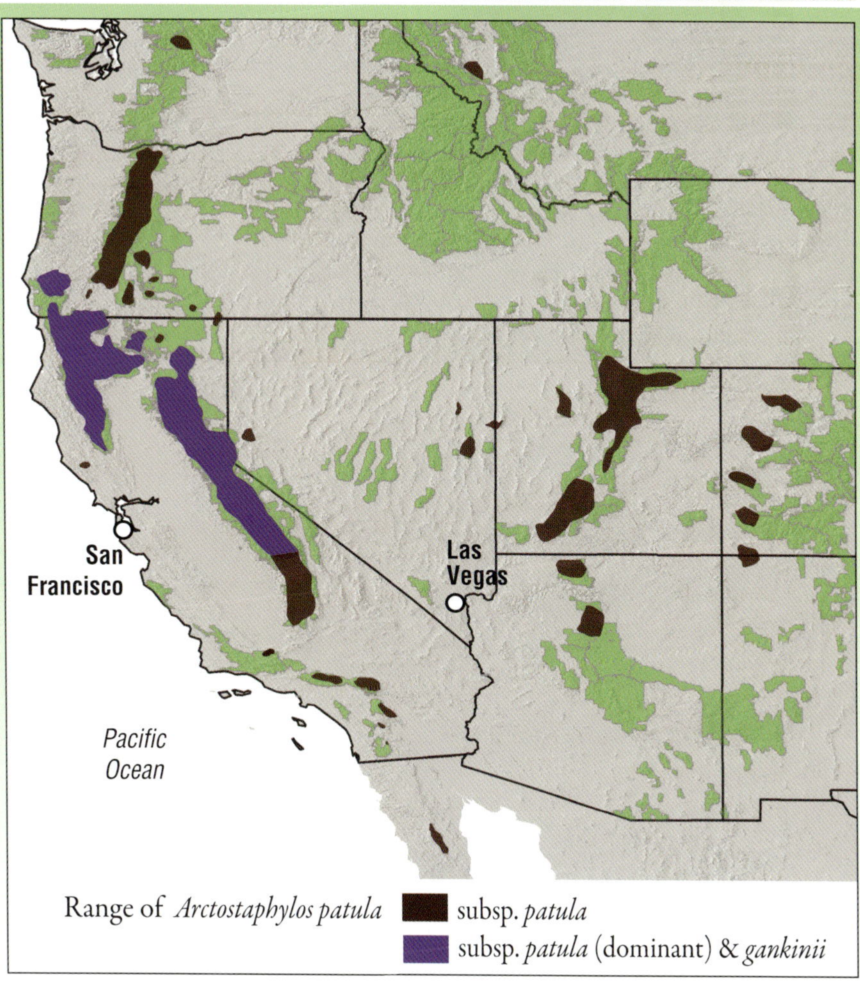

Pecho Manzanita

Arctostaphylos pechoensis

FORM: erect, 1-5 m STEM: twig short- and long-white-nonglandular-hairy LEAVES: overlapped; blade 2-5 cm, 1-2.5 cm wide, oblong-ovate, green-glaucous, dull to ±shiny, puberulent, in age glabrous, base lobed, clasping, tip acute INFLORESCENCE: panicle, 1-4-branched; nascent inflorescence pendent, short- and long-white-nonglandular-hairy, axis 1-1.5 cm, >1 mm wide; bracts 8-15 mm, leaf-like, lance-linear, acuminate, glabrous; pedicel 5-10 mm, glabrous FRUIT: 8-12 mm wide, depressed-spheric, glabrous; stones generally fused HABITAT: shale outcrops, chaparral, closed-cone conifer forest; <500 m; Irish Hills in and around Montaña de Oro State Park, southwestern San Luis Obispo County REMARKS: differs from nearby *A. osoensis* as described; at the south end of Montaña del Oro State Park, it associates with *A. crustacea* subsp. *crustacea*; similar to *A. purissima* which grows 56 km south near Vandenberg Air Force Base.

Morro Bay
Los Osos
Avila Beach
Pacific Ocean
Range of *A. pechoensis*

Peninsula Manzanita
A. peninsularis subsp. *peninsularis*

FORM: mounding shrub, 1-3 m, basal burl present STEM: bark reddish, smooth LEAVES: elliptic, oblong-ovate, 2-4 cm long by 1-2.5 cm wide, obtuse or truncate at base; intensely white glaucous with thick waxy deposit, dull or matte, glabrous, smooth INFLORESCENCE: 3-7 branched compact panicle, nascent branches 5-7 mm long, stout, divergent,; bracts all scale-like, keeled, acuminate tip; pedicels glabrous FRUIT: 5-9 mm wide, glabrous; pulp thin; nutlets fused HABITAT: granite outcrops, forms chaparral at middle or higher elevations, 1000-2500 m; San Pedro Mártir; also disjunct in the Sierra San Borja, and isolated peaks between, notably Cerro San Luis in Sierra Calamajue REMARKS: completely glabrous leaves, with no hairs or glands of any kind; one of the most intensely pink flowers of any manzanita species, which stand out beautifully against its white-glaucous leaves; fairly low growing, mounding species; associates with *A. glauca* and *A. pringlei* subsp. *pringlei*, overlaps at the lower zone of *A. patula*.

Sierra Juarez Manzanita

A. peninsularis subsp. *juarezensis*

FORM: erect, single-stemmed shrub, 1-5 m, no burl HABITAT: granite outcrops, 1750-2000 m among Jeffrey pines and the Sierra Juarez pinyon pines; Laguna Hansen, Sierra Juarez, Baja California REMARKS: nearly identical to subsp. *peninsularis* in its leaf, flower, and fruit characters, also in having completely glabrous branchlets and inflorescence without hairs or glands; differs mainly in its non burled, erect, single stemmed habit compared to the low, mounding burl forming habit of subsp. *peninsularis*; occurs at elevation above where *A. glauca* grows in this area and within the upper limits of *A. pungens* (very common in the Sierra Juarez), also associates with the uncommon *A. pringlei* subsp. *pringlei*; considered by some to be one of the most beautiful of all manzanita species with intensely white-glaucous leaves contrasting with its deep rose-colored flowers and red-orange trunks.

Range of *A. peninsularis*
subsp. *peninsularis*
subsp. *juarezensis*

Santa Margarita Manzanita
Arctostaphylos pilosula

FORM: erect, 1-5 m STEM: twig densely long-nonglandular-bristly-hairy LEAVES: erect; blade 1-3 cm, 1-2 cm wide, narrowly elliptic to round-ovate, dark green to gray-glaucous, dull, glabrous, base wedge-shaped, truncate to ±lobed, tip obtuse or acute INFLORESCENCE: +- raceme, 0-1-branched; nascent inflorescence pendent, long-nonglandular-hairy, axis 1-2 cm, >1 mm wide,; bracts ascending, 8--15 mm, lanceolate, leaf-like, green, persistent; pedicel 2-5 mm, glabrous FRUIT: 8-10 mm wide, depressed-spheric, glabrous; stones free HABITAT: shale outcrops, slopes, chaparral; 30-1250 m; southern Santa Lucia, La Panza Range, to hills around Pismo Beach, for example, Pozo Summit in the La Panza Range REMARKS: distinguished by its dense long bristly hairs; unique, long narrow ascending bracts on more or less unbranched (racemose) panicles; fruit is smooth and lacking hairs; new leaves often have a purple tint; interior populations have gray-glaucous colored leaves while the coastal populations have dark green colored leaves; however, this leaf color variation, and other characteristics, vary among individuals along a coast to interior gradient; originally described from the La Panza Range east of Santa Margarita; species needs further study to determine if the coastal and interior populations should be elevated to subspecies.

Pinkbract Manzanita

A. pringlei subsp. *pringlei*

FORM: erect, 1-5 m STEM: twig (and inflorescence axis) densely short-glandular-hairy LEAVES: erect; blade 2-5 cm, 1-4 cm wide, ovate, elliptic, or ±round, gray-glaucous, glandular-hairy, base rounded, truncate, or ±lobed, tip abruptly soft-pointed INFLORESCENCE: +- raceme, 0-1-branched; inflorescence forming just before flower (unlike most or rest of genus), pendent, axis 1-1.5 cm, >1 mm wide; bracts 6-10 mm, leaf-like, lanceolate, acute, deep pink, glandular-hairy, deciduous after flower; pedicel 5-15 mm, glandular-hairy FRUIT: 6-12 mm wide, spheric, glandular-hairy, sticky; stones separable HABITAT: rocky slopes, open conifer forest; 1000-2400 m; primarily in Nevada and Arizona, extends into Baja California, for example, Virgin Mountains, Nevada and Chiricahua Mountains of southeastern Arizona REMARKS: large, wide gray-green leaves are covered with dense, short glandular hairs, making them sticky to the touch; these dense glandular hairs also cover the inflorescence and fruit; fruit with separate nutlets; unique among *Arctostaphylos* are the deciduous pink bracts and the late development of its immature inflorescence which emerge shortly before the flowers bloom; all other species produce the dormant immature inflorescences during the previous spring or summer and typically persist over the dry season until spring rains trigger blooming depending on elevation; subsp. *pringelei* does not occur in California but appears in the Peninsular Ranges of Baja California, both in the Sierra Juarez south of the border and in the Sierra San Pedro Mártir.

California Pinkbract Manzanita
A. pringlei subsp. *drupacea*

FRUIT: stones fused HABITAT: rocky slopes, open conifer forest; 1250-2250 m, San Bernardino Mountains, Peninsular Ranges; restricted to Southern California and parts of Baja California, for example, Barton Flats, San Bernardino National Forest and Cuyamaca Peak in Cuyamaca Rancho State Park REMARKS: differs from subsp. *pringlei* in having consistently fused nutlets into a solid stone; the two subspecies ranges can overlap in Baja California.

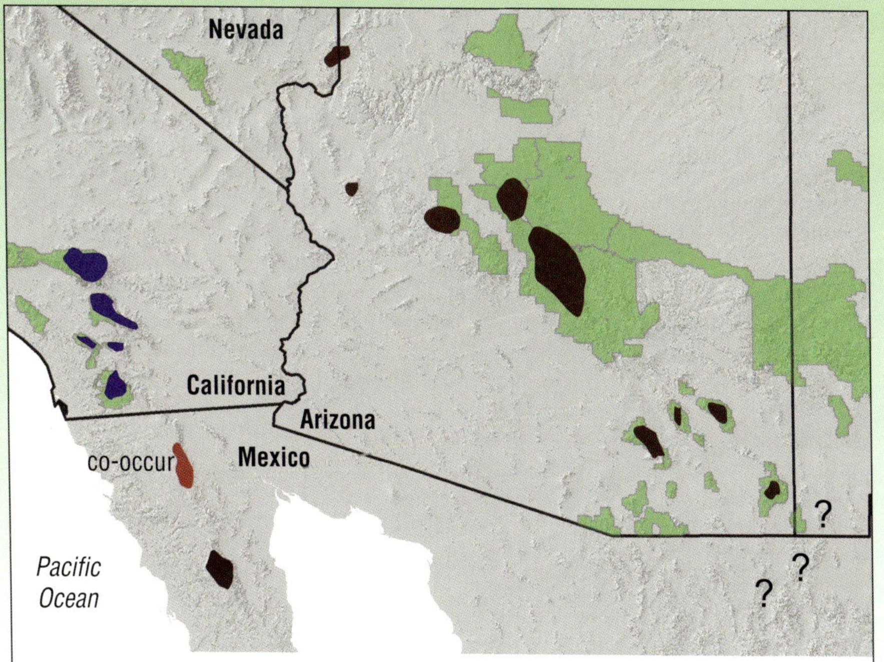

Range of *Arctostaphylos pringlei*

Sandmat Manzanita

Arctostaphylos pumila

FORM: prostrate to mounded, rarely erect, 0.1-1 m STEM: ±red, persistent, ±shredding; twig (and nascent inflorescence axis) short-gray-nonglandular-hairy LEAVES: spreading; blade 1-2 cm, 0.5-1.5 cm wide, narrowly obovate to oblanceolate; lower leaf gray-tomentose in youth, sparsely so in age, upper leaf convex, dark green, ±shiny, sparsely puberulent, lacking stomata; base wedge-shaped, tip obtuse, margin entire, cupped, tinged with red INFLORESCENCE: +- raceme, 0-1-branched; nascent inflorescence pendent, axis 0.3-0.5 cm, >1 mm wide; bracts 2-3 mm, scale-like, acute to sharp-pointed, glabrous or nonglandular-hairy; pedicel 3-4 mm, nonglandular-hairy FRUIT: 5-6 mm wide, spheric, sparsely nonglandular-hairy to glabrous; stones free HABITAT: sandy soils in chaparral, oak woodland; <200 m; surrounding Monterey Bay, for example, near Laguna Seca Golf Course, at end of York Rd.; Fort Ord National Monument REMARKS: prostrate to mounding, rooting along the ground and forming large, circular mounds; branchlets covered in short, gray, non-glandular hairs, similar to *A. tomentosa*; at Fort Ord, it occurs with *A. tomentosa, A. montereyensis, A. pajaroensis,* and *A. hookeri,* with occasional, apparent hybrids.

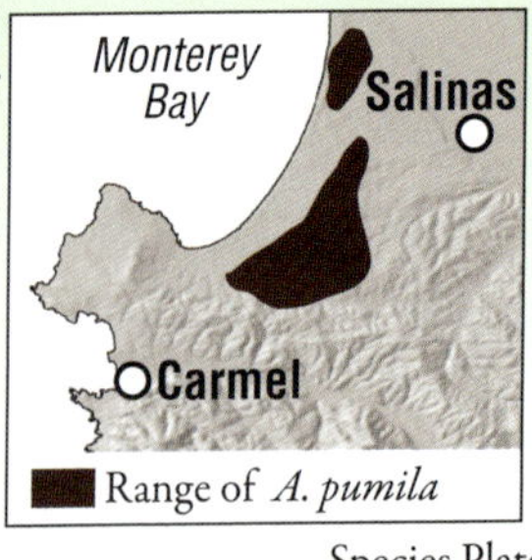

Mexican Manzanita

Arctostaphylos pungens

FORM: mounded to erect, 1-3 m STEM: twig (and nascent inflorescence axis) sparsely short-nonglandular-hairy LEAVES: erect; blade 1.5-4 cm, 1-1.8 cm wide, elliptic to lance-elliptic, bright or dark green, shiny, short-hairy, glabrous in age, base obtuse to wedge-shaped, occasionally rounded, tip acute INFLORESCENCE: +- raceme, 0-1-branched; nascent inflorescence pendent, club-like, axis 0.5-1.5 cm, > 1 mm wide; bracts 2-4 mm, scale-like, ovate-deltate, acuminate, recurved, glabrous; pedicel 5-10 mm, glabrous FRUIT: 5-8 mm wide, depressed-spheric, glabrous; stones free (rarely fused) HABITAT: rocky slopes, ridges, chaparral, conifer forest; 180-2300 m; ranges from Oaxaca, in southern Mexico to the southwestern United States and central coast region of California; for example, Peninsular and Transverse Ranges up to the Mount Pinos area REMARKS: unique immature inflorescence which is typically a tightly compacted raceme with scale-like bracts spreading in a strongly recurved manner into a distinctive "club" shape; may have both erect and prostrate or mounding forms, often growing in the same area one of the three most widely distributed manzanitas including *Arctostaphylos uva-ursi* and *A. patula*, Along central coast of California, it is found in three disjunct populations (Pinnacles National Monument, interior of the Santa Lucia Mountains, and San Benito County), and these populations show more variation than all of the southern populations; each of these populations were once given separate species names; reportedly also found in southern Baja.

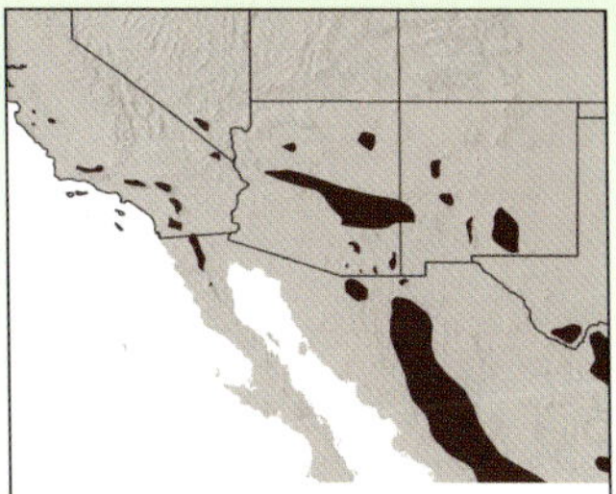

Range of *A. pungens*

La Purissima Manzanita

A. purissima subsp. *purissima*

FORM: erect, 1-4 m (occasionally prostrate) STEM: twig densely short- and sparsely long-nonglandular-hairy LEAVES: strongly overlapped; blade 1-2.5 cm, 1-2.5 cm wide, round-ovate to ±round, bright green, shiny, glabrous, base lobed, clasping, tip abruptly soft-pointed INFLORESCENCE: +- raceme, 0-2-branched; nascent inflorescence pendent, axis 0.5-1 cm, > 1 mm wide, densely nonglandular-hairy; bracts overlapped, 5-8 mm, leaf-like, ovate to lance-ovate, glabrous; pedicel 3-5 mm, ovary glabrous FRUIT: 5-8mm wide, depressed-spheric, glabrous; stones free HABITAT: sandstone outcrops, sandy soils, chaparral; <300 m; w Santa Barbara Co., for example, around Vandenberg Air Force Base REMARKS: short-dense and long-sparse, nonglandular hairs (similar to *A. pechoensis*); simple raceme or 1-2 branched panicles with wide, leafy bracts; smaller, depressed-spheric fruit with nutlets that are separable at Point Sal, prostrate forms of *A. purissima* occur, retaining their prostrate habit even when grown in less exposed sites; also similar to *A. viridissima* of Santa Cruz Island.

Range of *A. purissima*
subsp. *purissima*

Globose Manzanita

A. purissima subsp. *globosa*

FORM: mounding to erect, 1-3 m STEM: twig with dense short hairs and sparsely long-glandular-hairs INFLORESCENCE bracts usually densely ciliate with some gland-tipped hairs; pubescent pedicels, ovary glabrate FRUIT: 4-8 mm wide, globose, glabrous; nutlet frequently fused, HABITAT: sandstone outcrops, sandy soils, chaparral; <300 m; w Santa Barbara Co. REMARKS: glandular-hairy on twigs and inflorescences, globose fruit frequently fused; south of Hwy 246, plants occur in maritime chaparral, associates include *A. glandulosa* subsp. *glandulosa, Pickeringia montana, Vaccinium ovatum,* and *Quercus agrifolia*; individuals are often intermixed with coastal sage scrub or oak woodland.

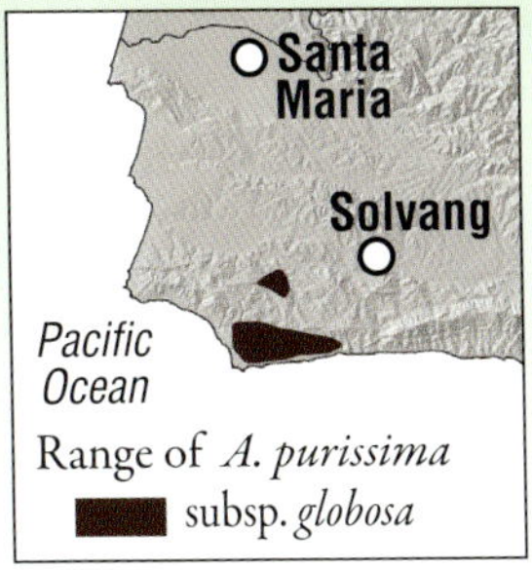

Rainbow Manzanita

Arctostaphylos rainbowensis

FORM: erect, 1-4 m; burled STEM: twig minutely glandular, appearing glabrous LEAVES: erect; blade 3.5-5 cm, 2-3.5 cm wide, elliptic-ovate, light green to gray-glaucous, dull, glabrous, base rounded, tip obtuse INFLORESCENCE: panicle, 4-10-branched; nascent inflorescence pendent, axis 2-3 cm, > 1 mm wide, sparsely short-glandular-hairy; bracts appressed, 2-4 mm, scale-like, deltoid to awl-shaped, glabrous; pedicel 5-10 mm, glabrous FRUIT: 8-12 mm wide, spheric, white-glaucous, dark brown, purple-tinged, glabrous; stones fused HABITAT: granitic outcrops, chaparral; 150-800 m; Agua Tibia Mountains and as far west as the Santa Margarita Mountains REMARKS: glabrous fruit with an unusual purple tinge and a whitish bloom and a solid stone; large basal burl; its branchlets and inflorescence have sparse glandular hairs; large, descending panicle with well-spaced, non-overlapping bracts; originally considered a San Diego county population of *A. peninsularis*; one of the only burl-forming diploid species (others being *A. patula, A. incognita,* and *A. rudis*).

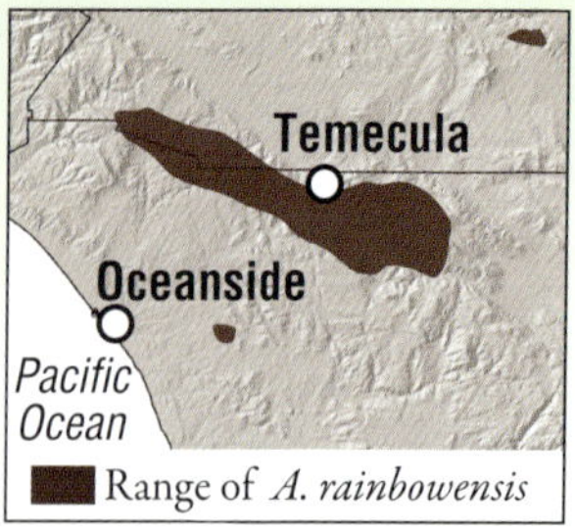

Refugio Manzanita

Arctostaphylos refugioensis

FORM: erect, 2-4 m STEM: twig (and immature inflorescence axis) densely short- nonglandular- and -long-glandular-hairy LEAVES: overlapped; petiole <2 mm; blade 3-4.5 cm, 2-3 cm wide, wide- to oblong-ovate, glaucous, dull, glabrous or midrib ±hairy, base lobed, clasping, tip acute INFLORESCENCE: panicle, 5-10-branched; nascent inflorescence pendent, axis 2-3 cm, > 1 mm wide; bracts 4-10 mm, leaf-like, ovate to widely elliptic, glandular-hairy; pedicel 7- 9 mm, glandular-hairy FRUIT: 10-15 mm wide, spheric, glabrous; stones fused HABITAT: sandstone outcrops, chaparral; 300-820 m; 24 km western Santa Barbara Co., for example, on a ridge above Refugio Beach in the Santa Ynez Mountains REMARKS: dense, short and sparse long, non-glandular hairs and bright-green leaves; differs from all other auriculate-leaved species (except *A. gabilanensis*) in having large, spheric fruit with a solid stone; nearest auriculate manzanita species is *A. purissima* subsp. *globosa* on the western end of the Santa Ynez Mountain; the southernmost of the auriculate-leaved manzanitas on mainland California.

King's Mountain Manzanita

Arctostaphylos regismontana

FORM: erect, 2-5 m STEM: twig glandular-hairy LEAVES: overlapped; blade 3-6 cm, 2-3 cm wide, oblong-ovate, boat-shaped, pale green, dull, glandular-sticky-hairy, glabrous in age, base lobed, clasping, tip acute INFLORESCENCE: panicle, 4-6-branched; nascent inflorescence pendent, glandular-hairy, axis 1.5-2 cm, >1 mm wide; bracts 5-12 mm, leaf-like, narrowly lance-oblong; pedicel 6-10 mm, glandular-hairy FRUIT: 6-8 mm wide, depressed-spheric, sticky; stones free HABITAT: granite, and sandstone outcrops, edge of conifer forest, chaparral; 150-780 m; King's Mt. Rd. & Skyline Blvd. San Mateo Co. REMARKS: large, tree-like; leaves are somewhat grayish and densely glandular; panicles are thicker than those of *A. andersonii* and much more glandular, found between the ranges of the auriculate-leaved *A. andersonii* to the south and *A. montaraensis* to the north; although occupying a slightly larger overall range than *A. montaraensis*, is much less common within this range.

Sand Mesa Manzanita

Arctostaphylos rudis

FORM: erect, 1-3 m; burl (occasionally not visible) STEM: old stem bark persistent, gray, shredding; twig (and nascent inflorescence axis) sparsely short-nonglandular-hairy LEAVES: erect; blade 1–3 cm, 1–2 m wide, elliptic, bright green, shiny, puberulent, in age glabrous, base wedge-shaped to rounded, tip acute INFLORESCENCE: +- raceme, 0-1-branched; nascent inflorescence pendent, sparsely short-nonglandular-hairy, axis 0.5-1 cm, >1 mm wide; bracts 2-6 mm, scale-like, deltate to acuminate, sharp-pointed, glabrous; pedicel 3-6 mm, glabrous FRUIT: 8-14 mm wide, depressed-spheric, glabrous; stones free HABITAT: sandy soils, chaparral; <380 m; s San Luis Obispo, n Santa Barbara, for example, Burton Mesa, Point Sal REMARKS: persistent shaggy bark typically with a basal burl; like *A. purissima*, has erect and low prostrate forms, often associates with *A. purissima*; on Point Sal Ridge, *A. rudis, A. purissima,* and *A. crustacea* subsp. *crustacea* are growing as low, spreading mats, perched high above the Pacific Ocean.

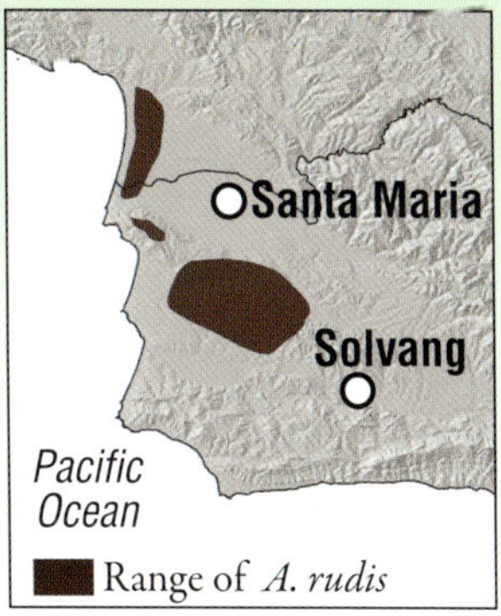

Coinleaf Manzanita

Arctostaphylos sensitiva

FORM: erect, 1-2 m STEM: twig densely short-nonglandular and long-glandular-hairy LEAVES: spreading; blade 1-2.2 cm, 0.8-1.8 cm wide, round to round-ovate, shiny, glabrous, midvein hairy, lower leaf light-green, upper dark-green, lacking stomata; base truncate to ±lobed, tip obtuse INFLORESCENCE: panicle, 4-8-branched; nascent inflorescence pendent, short- nonglandular- and -long-glandular-hairy, axis 0.5-1 cm, > 1 mm wide; bracts appressed, 0.5-2 mm, scale-like, deltate, sharp-pointed, glabrous; pedicel 3-5 mm, glabrous. FLOWER: sepals and petals 4-merous FRUIT: 3-4 mm wide, subcylindric, glabrous, mature splitting; stones free HABITAT: rocky sites, chaparral, closed-cone conifer forest, <600 m; Marin, San Mateo, and Santa Cruz counties from Mount Tamalpais and Bolinas Ridge to the Santa Cruz Mountains in the south; REMARKS: branchlets and inflorescences with short-dense, non-glandular hairs and long-sparse glandular hairs; leaves with different surfaces, shiny dark green on the upper surface (lacking stomata) while light green on the bottom surface with stomata; shares the distinction with *A. nummularia* in having four sepals and four-lobed corolla instead of these flower parts in fives; fruits fall early and are rarely seen (also *A. myrtifolia*); once considered same species as *A. nummularia*.

Bonny Doon Manzanita

Arctostaphylos silvicola

FORM: erect, 1-4 m STEM: twig sparsely to densely sofi-nonglandular-hairy LEAVES: erect; blade 1.5-3.5 cm, 1-1.5 cm wide, narrowly obovate or oblong-elliptic, gray, dull, appressed-canescent, in age glabrous, glaucous, base wedge-shaped, tip obtuse INFLORESCENCE: +- raceme, 0-1-branched; nascent inflorescence pendent, axis 0.5-1 cm, > 1 mm wide; bracts 5-12 mm, leaf-like, lanceolate, canescent to glabrous; pedicel 5-7 mm, sparsely nonglandular-hairy or glabrous FRUIT: 6-12 mm wide, spheric, glabrous; stones free HABITAT: weathered sandstone soils in chaparral, conifer forest; <600 m; Santa Margarita sandstone at Bonny Doon, Mount Herman and Quail Hollow REMARKS: differs from *A. canescens* (occurs ~32 km away on Loma Prieta Peak) primarily in having smaller, narrow leaves, consistently white flowers, glabrous ovaries and large, smooth fruit not covered in fine hair; at Bonny Doon, it associates with *A. sensitiva, A. crustacea* subsp. *crinita,* and *A. andersonii.*

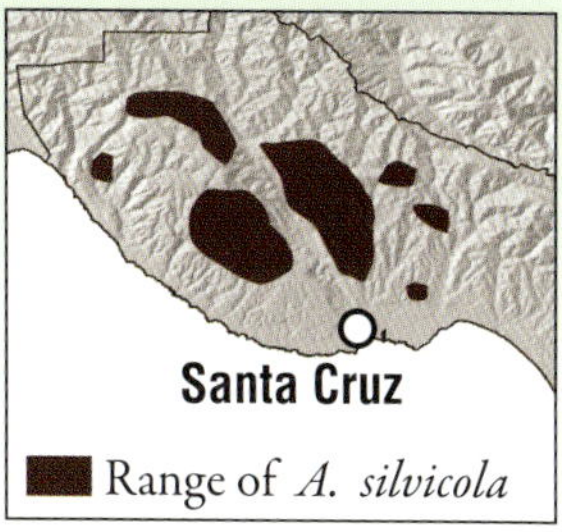

Stanford's Manzanita

A. stanfordiana subsp. *stanfordiana*

FORM: erect, 1-3 m STEM: twig glabrous LEAVES: glabrous, petiole 8-12 mm, 1.5-2.5 cm wide, elliptic, oblong, or oblanceolate, glabrous, bright green ±shiny, base wedge-shaped, tip abruptly soft-pointed INFLORESCENCE: panicle, 3-5-branched; nascent inflorescence erect to pendent, axis 2-2.5 cm, < 1 mm wide; bracts 1-1.5 mm, scale-like, ovate to deltate, acuminate, generally glabrous; pedicel 5-10 mm, glabrous FRUIT: 6-8 mm wide, depressed-spheric, glabrous; stones free HABITAT: chaparral in North Coast Range; 150-1000 m; Humboldt, Mendocino, Sonoma, and Napa rocky uplands, e.g., Mount Saint Helena in Napa County REMARKS: delicate overall appearance; distinguished by thin-branched inflorescence with minute scale-like bracts framing widely scattered buds; widespread where it associates with *A. manzanita*, *A. glandulosa*, and *A. canescens*.

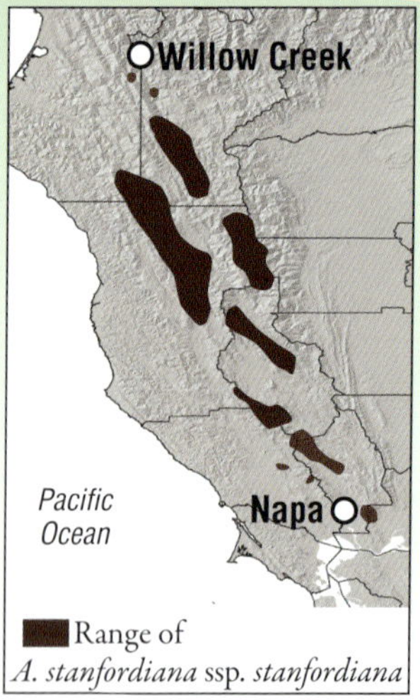

Rincon Ridge Manzanita

A. stanfordiana subsp. *decumbens*

FORM: prostrate to mounded, 0.1-1 m LEAVES: petiole +- 4-8 mm HABITAT: chaparral; 100 m; Rincon Ridge north of Santa Rosa REMARKS: similar to *A. stanfordiana*, but growing prostate to slightly mounded occurs in red rhyolite soils and retains its low spreading habit even when grown on other soils; at Rincon Ridge it associates with *A. canescens* subsp. *sonomensis*, *A. manzanita* subsp. *manzanita*, and *Ceanothus confusus*; local endemic.

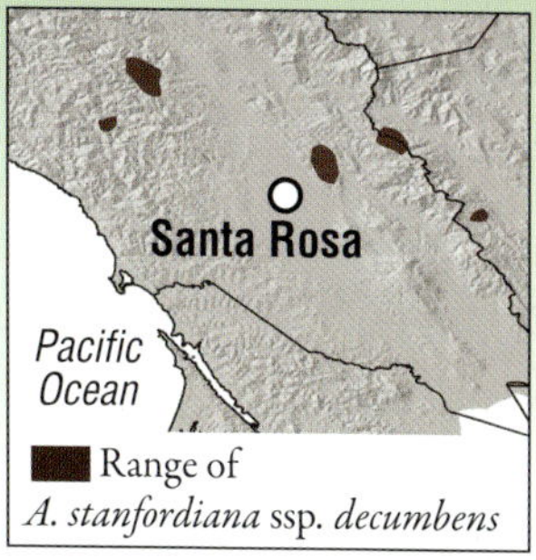

Raiche's Manzanita

A. stanfordiana subsp. *raichei*

STEM: twig short-glandular-hairy LEAVES: papillate, petiole +- 4-8 mm HABITAT: chaparral; 400-945 m; Mill Creek Rd., South Cow Mountain Rec. Area, east of Ukiah REMARKS: new growth branchlets (and sometimes the inflorescences) have glandular hairs, and sometimes the leaf blades have glands on the lower surfaces and the primary veins; intermediate between the non-glandular *A. stanfordiana* subsp. *stanfordiana* and the densely glandular *A. hispidula* of Del Norte County; preference to sandstone barrens and serpentine soils; in the vicinity of Cloverdale, occurs mixed with subsp. *stanfordiana* and intermediate individuals occur.

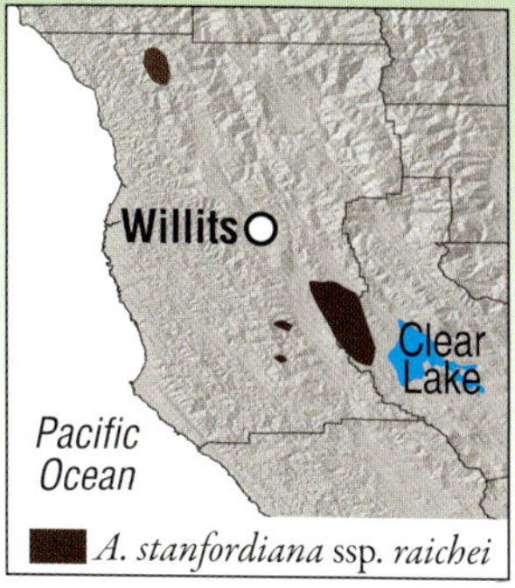

Woolyleaf Manzanita

A. tomentosa subsp. *tomentosa*

FORM: erect, 1-3 m; burl prominent STEM: twig sparsely short-tomentose; old stems gray, shreddy LEAVES: below densely tomentose; above dark to bright green, ±shiny, lacking stomata; spreading; blade 2-5 cm, 1.5-2.5 cm wide, oblong-ovate to lance-oblong, abaxially dull, base truncate to ± lobed, tip acute, occasionally toothed INFLORESCENCE: panicle, 2-8-branched; nascent inflorescence pendent, axis 1-2.5 cm, > 1 mm wide; bracts 8-15 mm, leaf-like, lanceolate; pedicel 2-5 mm FRUIT: 6-10mm wide, depressed-spheric, nonglandular; stones free HABITAT: chaparral, closed-cone conifer forest; 50-500 m; Monterey Peninsula REMARKS: distinguished from the other subspecies by the short, wooly, non-glandular hairs on the lower surface of the leaves and branchlets; lacks the typical smooth red bark found on most other manzanita species, instead presenting gray, shreddy bark on its lower stems; one of three related burl-forming tetraploid manzanitas (also *A. glandulosa* & *A. crustacea*); *A. tomentosa* and *A. crustacea* were once part of the same complex (10 subspecies), later separated into two species, *A. tomentosa* with gray, shreddy bark and *A. crustacea* with smooth, red bark. All subspecies have shiny leaves on the upper surface and gray, tomentose lower surfaces with cupped margins, panicles with leafy bracts, and shreddy bark on trunks with large prominent burls; subspecies differ mainly in hair patterns.

Glandular Woolyleaf Manzanita

A. tomentosa subsp. *bracteosa*

STEM: twig densely glandular-bristly; LEAVES: above sparsely glandular-hairy, papillate, scabrous or +- glabrous shiny; below- dull-green; sparsely glandular-hairy; blade 2-5 cm, 1.5-2.5 cm wide, oblong-ovate to lance-oblong, base truncate to ±lobed, tip acute, occasionally toothed INFLORESCENCE: densely glandular-bristly FRUIT: glandular-hairy HABITAT: chaparral, closed-cone conifer forest; 10-300 m; Jacks Peak and Fort Ord, Monterey Co. REMARKS: differs from subsp. *tomentosa* in having dense, long glandular hairs on the branchlets and immature inflorescence, and the leaves have sparse, glandular hairs on the lower surface instead of dense, wooly non-glandular hair; associates with subsp. *hebeclada* and *A. hookeri* subsp. *hookeri* on Jacks Peak.

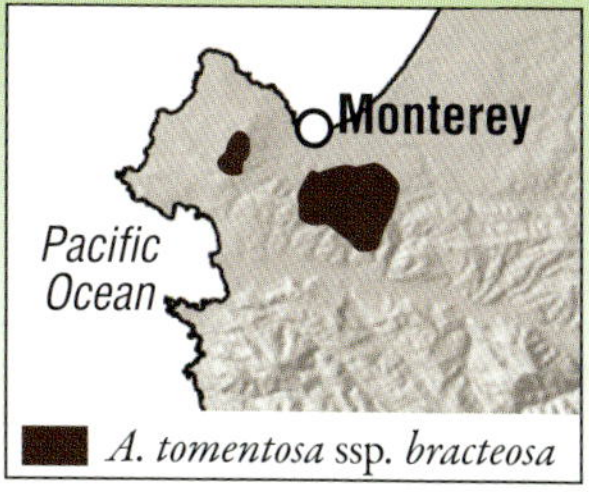

Dacite Manzanita

A. tomentosa subsp. *daciticola*

STEM: twig sparsely sparse-short- and long-white-nonglandular-hairy; INFLORESCENCE: sparse-short- and long-white-nonglandular-hairy HABITAT: chaparral; 200-300 m; small area west of San Luis Obispo on and near Hollister Peak REMARKS: has both short and long non-glandular hairs; only subspecies of A. *tomentosa* that does not occur on the Monterey Peninsula; near Los Osos State Park; *A. tomentosa* subsp. *daciticola* associates with *A. osoensis*.

Explorer's Manzanita

A. tomentosa subsp. *hebeclada*

STEM: twig sparsely short-nonglandular-hairy LEAVES: lower leaf surface glabrous FRUIT: 6-10 mm wide, depressed- spheric, nonglandular or glandular-hairy HABITAT: chaparral, closed-cone conifer forest, 100-300 m; Jacks Peak, Monterey Peninsula REMARKS: lower surface of leaf is glabrous (subsp. *tomentosa* has dense, wooly hairs); branchlets and inflorescence have sparse, short, nonglandular hairs; differs from subsp. *bracteosa* in lacking the long glandular hairs; similar to *A. crustacea* subsp. *rosei* in having leaves with different surfaces that are glabrous on the both surfaces but dark green above and light green below, differs in having gray-shreddy bark (*A. crustacea* subsp. *rosei* with smooth red bark); three of the four subspecies are found on the Monterey Peninsula, (subsp. *tomentosa, bracteosa,* and *hebeclada*).

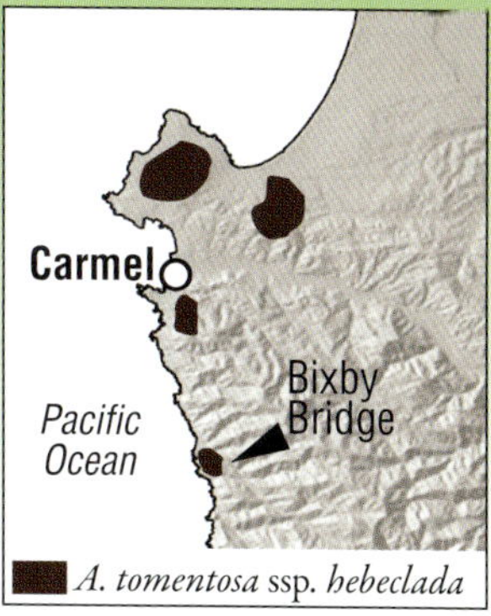

Bearberry or Kinnikinnick

A. uva-ursi subsp. *uva-ursi*

FORM: generally prostrate mats, rarely mounded, 0.1-0.5m; lacking a basal burl except occasionally small burls that occur on spreading branches STEM: twig (and nascent inflorescence axis) sparsely short-nonglandular-hairy, but some forms occasionally long-nonglandular- and/or short-glandular-hairy LEAVES: blades 1-2.5 cm, 0.5-1.5 cm wide, oblanceolate to obovate, occasionally narrowly elliptic; surfaces differ, below light-green, shiny, sparsely puberlent; above dark-green, shiny, lacking stomata; often cupped; spreading INFLORESCENCE: +-raceme, 0-1-branched; nascent inflorescence pendent, axis 0.3-1 cm, >1 mm wide; bracts 2-6 mm, scale-like, narrowly deltate, acuminate, glabrous; pedicel 2-4 mm, glabrous. FRUIT: 6-12 mm wide, red, spheric, glabrous; stones free HABITAT: rocky outcrops, slopes, stabilized dunes, closed-cone conifer forest, grassy coastal headlands, tundra, subalpine forest; w. N.A., <100 m or >2400 m; for example, upper Convict Lake drainage, Sierra Nevada, and Lanphere Dunes, Humboldt Co. REMARKS: low-growing with different leaf surfaces and bright red fruit with small stones embedded in pulpy tissue; may be bird dispersed; divided into eight named forms, five in California (*coactilis, adenotricha, marinensis, suborbiculata, leobreweri*), which formerly were considered varieties or subspecies; controversy surrounds the age of *A. uva-ursi* – is it a relatively young species or relatively old?

Guatemala Bearberry

A. uva-ursi subsp. *cratericola*

FORM: lacking burls STEM: long-glandular hairy to short non-glandular hairy LEAVES: thick, narrowly elliptic to oblanceolate, not bluntly rounded at tip INFLORESCENCE: pedicels often with long, glandular hairs HABITAT: Found on rocky high volcanic mountains and the limestone massif of the Sierra Cuchamatanes in Guatemala; REMARKS: Remarkable disjunction in the distribution of this subspecies. Apparently absent from all of Mexico, nearest related populations to the north in Northern California, northeastern Arizona, and northern New Mexico. Relatively few isolated populations in Guatemala. Needs further study, subsp. *cratericola* should be considered globally threatened.

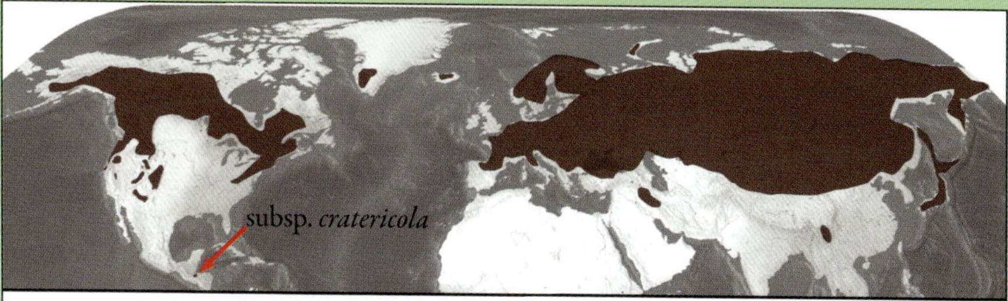

World Distribution of *Arctostapylos uva-ursi*

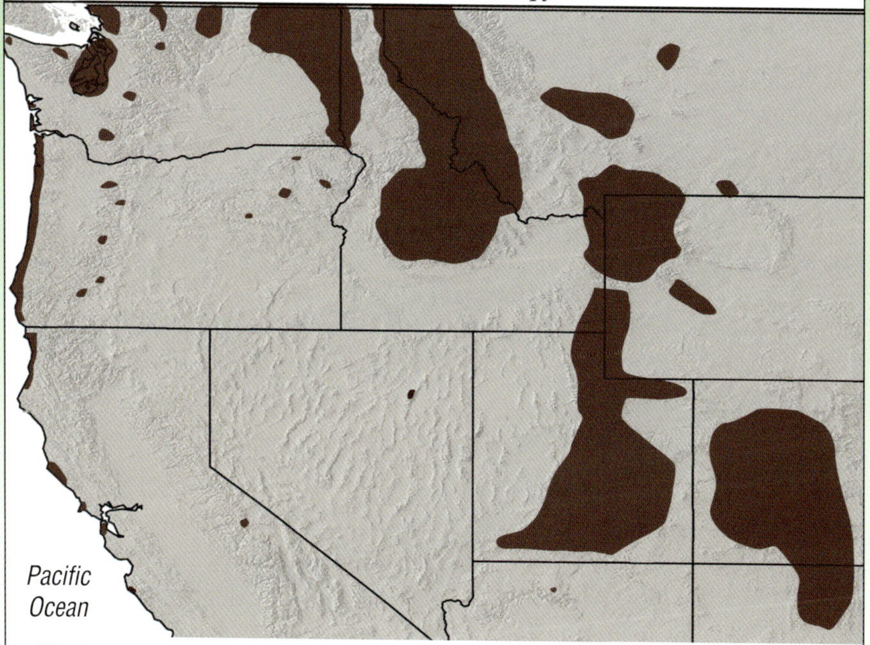

Range of *Arctostaphylos uva-ursi* in western North America

Marin Manzanita

Arctostaphylos virgata

FORM: erect, 1-5 m STEM: twig with short- and long-glandular-hairy LEAVES: ascending, overlapped; blade 3-5cm, 1-2.5cm wide, narrowly oblong-ovate to lance-oblong, sparsely glandular-hairy, papillate, bright green, ± shiny, base truncate to ±lobed, tip acute INFLORESCENCE: +- raceme, 0-1-branched; nascent inflorescence pendent, axis 1.5-2 cm, > 1 mm wide; bracts persistent, 8-20 mm, leaf-like, lance-linear, acuminate, long-glandular-hairy, green; pedicel 3-8 mm, glandular-hairy FRUIT: 6-8mm wide, depressed-spheric, glandular-hairy, sticky; stones free HABITAT: sandstone, granite outcrops in chaparral, conifer forest; <500 m; Marin Co., e.g. road to Point Reyes Lighthouse near the top of Inverness Ridge REMARKS: shiny, bright-green leaves that are "willow-like," being narrow and pointed; inflorescence branchlets are covered with dense, short and long, glandular hairs, sticky to the touch; limited distribution in Marin County on Inverness Ridge and Point Reyes, on Bolinas and Geronimo Ridge northwest of Mount Tamalpais, and on the south slope of Mt. Tamalpais near Muir Woods.

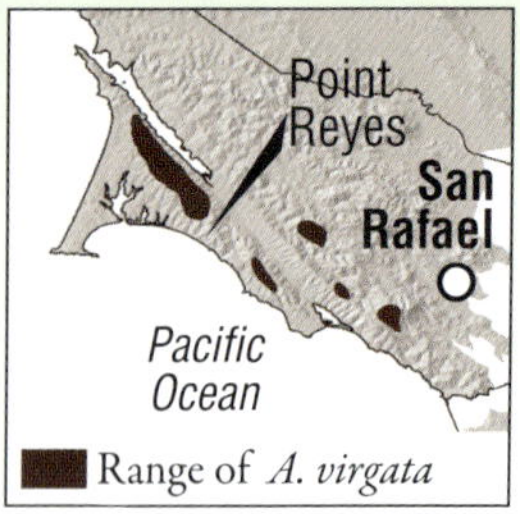

White-haired Manzanita

Arctostaphylos viridissima

FORM: erect, 1-4 m STEM: twig short- and long-white-nonglandular-hairy LEAVES: overlapped; blade 2-3.5 cm, 1.5-2.5 cm wide, narrowly to oblong-ovate, puberulent, glabrous in age, bright green, shiny, base truncate to lobed, ± clasping or not, tip abruptly soft-pointed INFLORESCENCE: +- raceme, 0-1-branched; nascent inflorescence pendent, axis 1-1.5 cm, > 1 mm wide; bracts 6-10 mm, leaf-like, lanceolate, acute, +- nonglandular-hairy; pedicel 2-3 mm, nonglandular-hairy FRUIT: 10-15 mm wide, ± spheric, sparsely nonglandular-hairy to glabrous; stones free HABITAT: shale outcrops, chaparral, closed-cone conifer forest; 100-550 m; eastern Santa Cruz Island between Pelican Bay and Scorpion Anchorage REMARKS: flower pedicels and ovaries that are densely hairy; long and short non-glandular hairs on branchlets; wide, rounded flowers, on short pedicels; erect, single stemmed, non-burl forming species, growing as scattered shrubs rather than forming large dense stand similar to *A. purissima* of the Burton Mesa north of Santa Barbara; found only on Santa Cruz Island, where it is the only auriculate or semi-auriculate leaved species.

Whiteleaf Manzanita

A. viscida subsp. *viscida*

FORM: erect, 1-4 m STEM: twig glabrous to sparsely short-nonglandular-hairy LEAVES: glabrous, entire, smooth, generally rounded; blade 2-5 cm, 2-4 cm wide, ovate to ±round, white-glaucous, dull, base rounded, truncate or ±lobed, tip abruptly soft-pointed INFLORESCENCE: panicle, 4-7-branched; nascent inflorescence erect to pendent, axis 1-3 cm, >1 mm wide, densely glandular-sticky-hairy; bracts appressed, 3-4 mm, scale-like, deltate, acute to acuminate, glaucous, sparsely short-nonglandular-hairy; pedicel 6-10 mm, glandular-hairy FRUIT: glabrous; 6-8 mm wide, depressed-spheric; stones free HABITAT: chaparral, forest; 30-2000 m; c Oregon, Sierra Nevada, and n Coast Range, for example, Whiskeytown Lake, Shasta Co.; Colfax, Placer Co. REMARKS: a glabrous species, lacking hairs or glands on its branchlets but sticky to the touch on its panicles; fruit are smooth, with no hairs or glands; differs from many of the manzanitas of the Coast Ranges which commonly have leaf-like bracts on their panicles; instead, *A. viscida* has small, scale-like bracts that are tightly appressed to the panicle species epithet "viscida" (meaning sticky) refers to the glands on the panicles, making them sticky to the touch; one of the most common, wide-ranging manzanitas in California.

Mariposa Manzanita

A. viscida subsp. *mariposa*

LEAVES: sparsely short- glandular-hairy, ciliate, papillate, scabrous erect STEM: twig glandular hairy INFLORESCENCE: ovary glandular hairy FRUIT: covered with glandular hairs HABITAT: openings in chaparral, forest; 400-2200 m. central and southern Sierra Nevada; for example, foothills of Yosemite National Park and across central Sierra Nevada foothills REMARKS: easily distinguished from subsp. *viscida*, which is glabrous, and non-glandular, while subsp. *mariposa* is densely glandular-hairy, also fruits and panicles covered with glandular hairs, branchlets and leaves are sticky and rough to the touch, compared to the completely smooth fruits, leaves and branchlets of subsp. *viscida*; fruits more glandular than subsp. *pulchella* of the interior North Coast Range when subsp. *viscida* and subsp. mariposa overlap, subsp. mariposa is usually found at higher elevations; SEE RANGE MAP NEXT PAGE.

Coast Whiteleaf Manzanita

A. viscida subsp. *pulchella*

STEM: twig glabrous to sparsely short-nonglandular-hairy LEAVES: glabrous; blade 2-5 cm, 2-4 cm wide, ovate to ± round, white-glaucous, dull, base rounded, truncate or ±lobed, tip abruptly soft-pointed INFLORESCENCE: ovary sticky-glandular, rough FRUIT: covered with glandular hairs, sticky, rough; 6-8mm wide, depressed-spheric; stones free HABITAT: openings in forest, chaparral; 150-1400 m, mostly on serpentine soils, growing in the North Coast Range from Sonoma County to southwestern Oregon, for example, The Cedars, Sonoma Co.; Mt. St. Helena, Lake Co. Framework: fruit are rough and glandular-sticky (subsp. *viscida* with completely smooth fruit); panicles are also densely glandular hairy, with leaves and branches that are glabrous, without hairs.

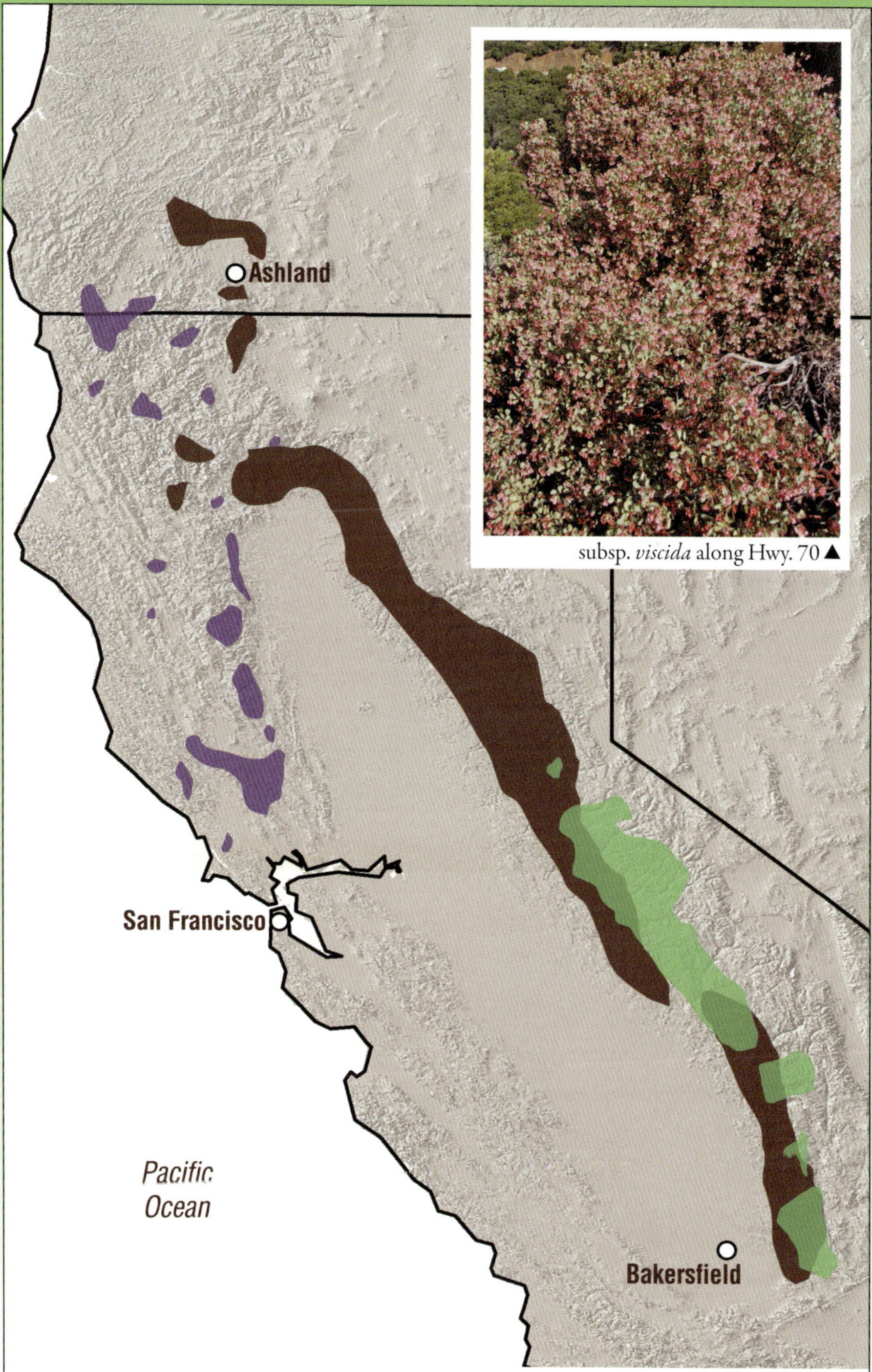

subsp. *viscida* along Hwy. 70 ▲

Range of *Arctostaphylos viscida*

subsp. *mariposa* subsp. *pulchella* subsp. *viscida*

Manzanita Destinations

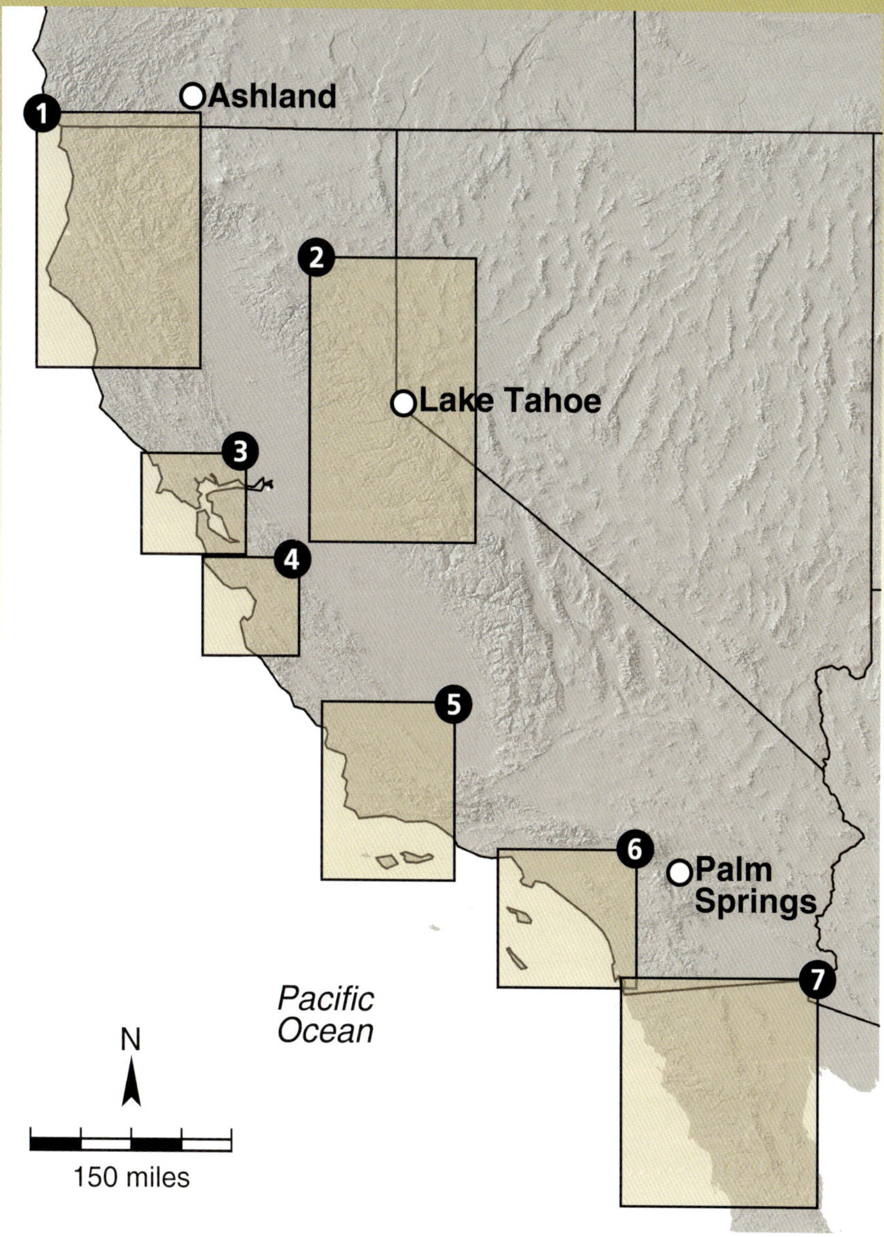

Manzanita Destinations

1. Northwest California
2. Central Sierra
3. San Francisco Bay
4. Monterey Bay
5. Central Coast
6. Southern California
7. Baja California

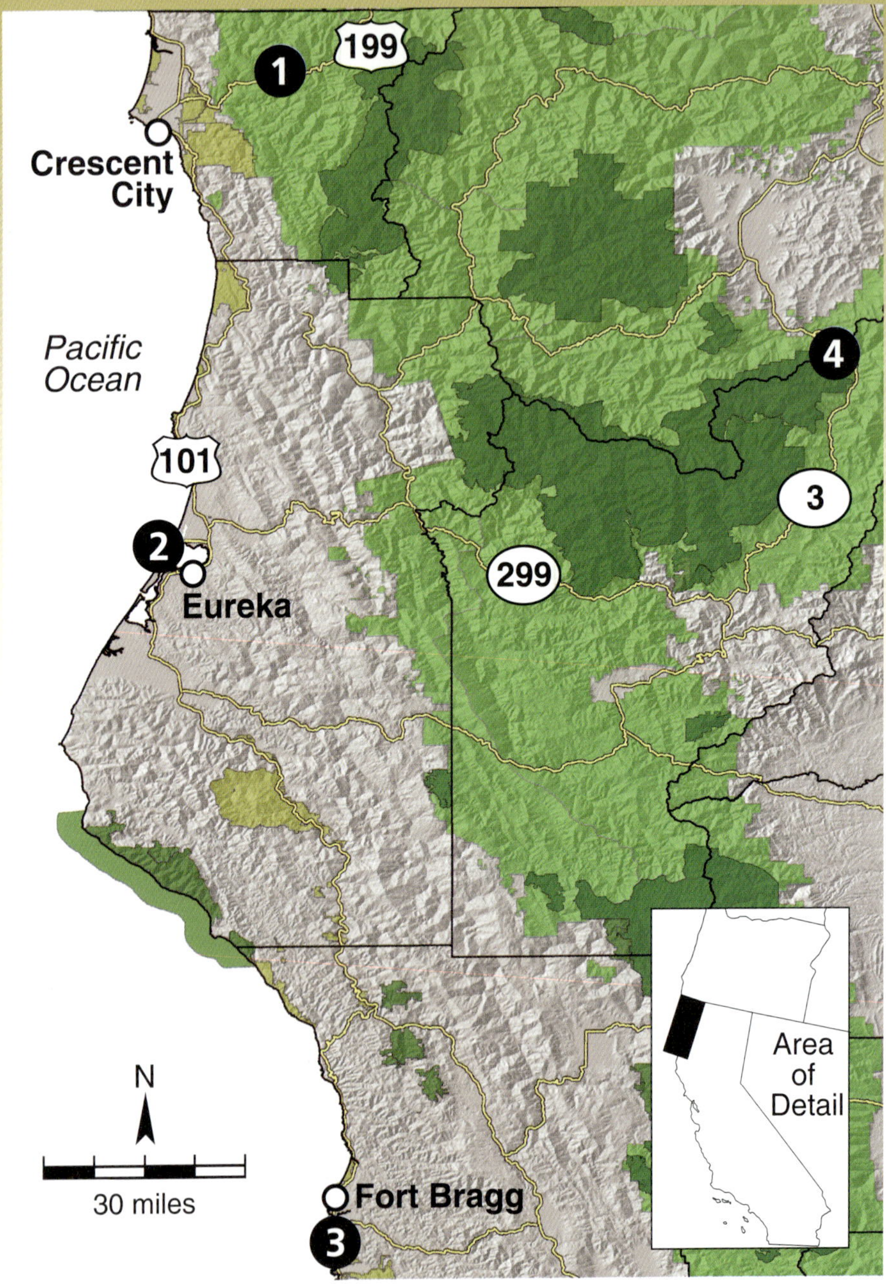

Manzanita Destinations: Northwest California

National Forest
Wilderness
State Park
Private Land

Manzanita Destination: Northwest California

1 Smith River National Rec. Area

Getting there: Travel Highway 199 northeast from Crescent City to Gasquet, California and visit the ranger station for more details.

Key species: *A. hispidula, A. nortensis, A. nevadensis* spp. *knightii*

Remarks: This region nurtures 70 endemic plants on the largest serpentine outcrop in North America - the Josephine Ophiolite.

2 Lanphere-Christensen Dunes

Getting there: From Highway 101 in Arcata, follow Samoa Blvd. west to the Ma-Le'l Dunes parking area. Visit friendsofthedunes.org to learn more.

Key species: *A. columbiana, A. uva-ursi*

Remarks: Just a few miles west of Arcata, California and a world apart from the majestic redwood forests, this dune complex is one of the most pristine on the Pacific Coast. Stunted forests attempt to hold shifting dunes while manzanita takes purchase in the understory.

3 Jug Handle State Reserve

Getting there: Located on the north coast of California about equal distance (five miles) between Fort Bragg and Mendocino in Mendocino County

Key species: *A. nummularia* spp. *mendocinoensis, A. columbiana*

Remarks: The best preserved show-place of coastal landscape evolution in the Northern Hemisphere is seen in the Ecological Staircases.

4 Scott Mountain Botanical Area

Getting there: From Weaverville, take Hwy 3 north for 53.4 miles, botanical area is on the north side of the Scott Mountain campground

Key species: *A. klamathensis, A. patula, A. nevadensis*

Remarks: The area represents an example of a mid-elevation(1200-1800 m) ultramafic forest type full of many rare plants including the Klamath manzanita and California pitcher plant.

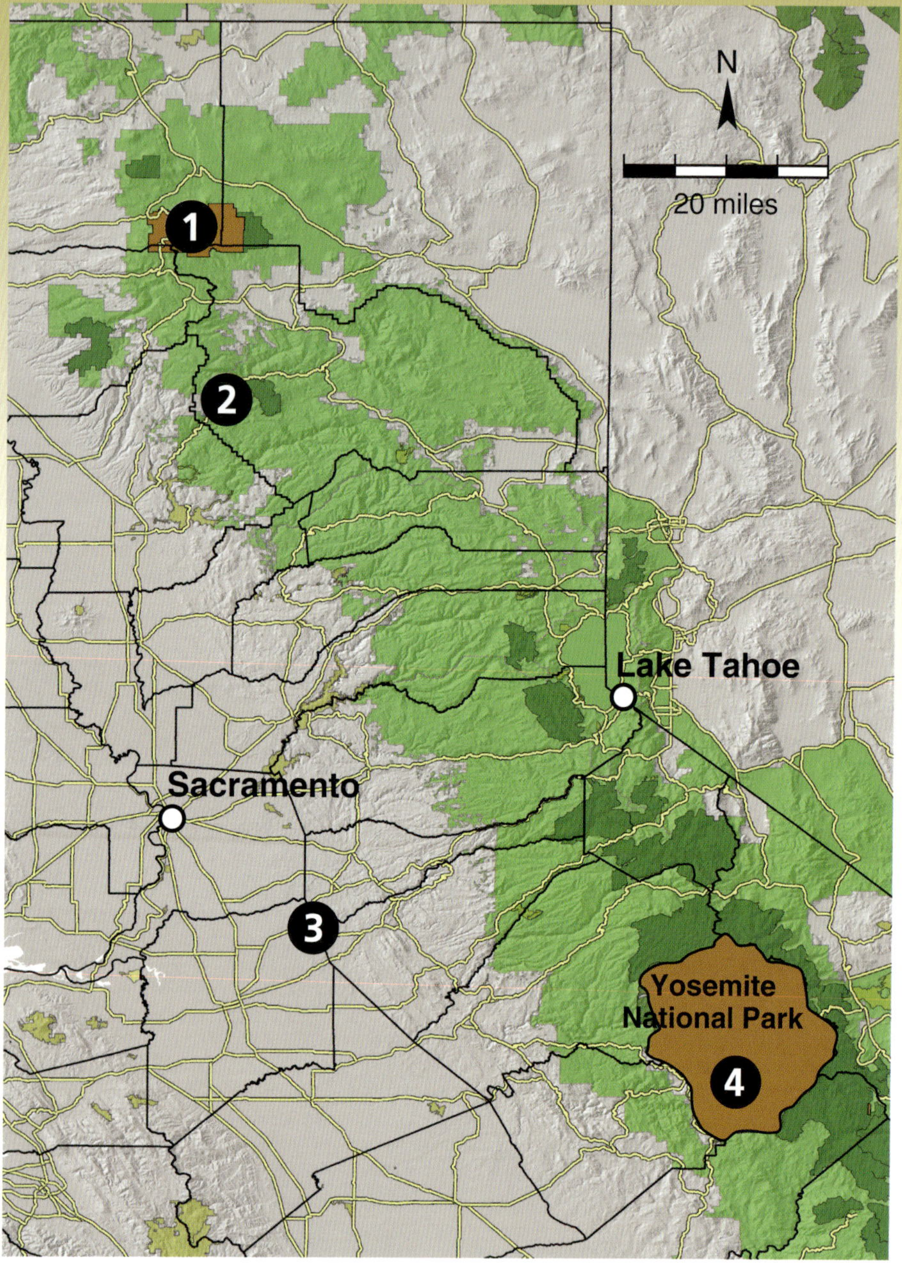

Manzanita Destination: Northern Sierra Nevada

National Forest
Wilderness
National Park
State Park
Private Land

Manzanita Destinations: Northern Sierra Nevada

1 Lassen Volcanic N. P.

Getting there: There are five separate entrances into the park, all leading to diverse landscapes; one road runs north-south through the park, leading into the high country; 165 miles from Sacramento and 180 miles from Reno

Key species: *A. nevadensis, A. patula*

Remarks: While Lassen N.P. is primarily known for its volcanic geology, the park boasts a rich diversity of plants, with over 700 flowering plant species.

2 Feather River Canyon

Getting there: Take highway 70 for any number of miles - accessed from near Oroville

Key species: *A. mewukka* ssp. *mewukka, A. viscida, A. manzanita*

Remarks: One of the most scenic highway drives in California, the route highlights the natural beauty of the Sierra Nevada. Numerous hiking options are offered along the route, including the Pacific Crest Trail.

3 Highway 88

Getting there: From Lodi, travel 15 miles northeast on Highway 88 to the junction with Buena Vista Road

Key species: *A. myrtifolia, A. viscida*

Remarks: Species are growing along a brief section of Hwy 88 just east of Buena Vista Rd.; also along Buena Vista Rd. just south of Hwy 88 with *A. viscida* subsp. *viscida, A. manzanita* subsp. *manzanita*, along with hybrids between *A. mytifolia* and *A. viscida.*

4 Yosemite National Park

Getting there: 195-miles from San Francisco or 315-miles from Los Angeles; travel Tioga Rd into the high country

Key species: *A. nevadensis, A. patula, A. viscida* ssp. *mariposa, A. viscida* ssp. *viscida*

Remarks: By traveling from the lower elevations to the upper reaches of the peaks, vegetation changes from oak woodlands to chaparral scrublands to lower montane to upper montane to subalpine to alpine.

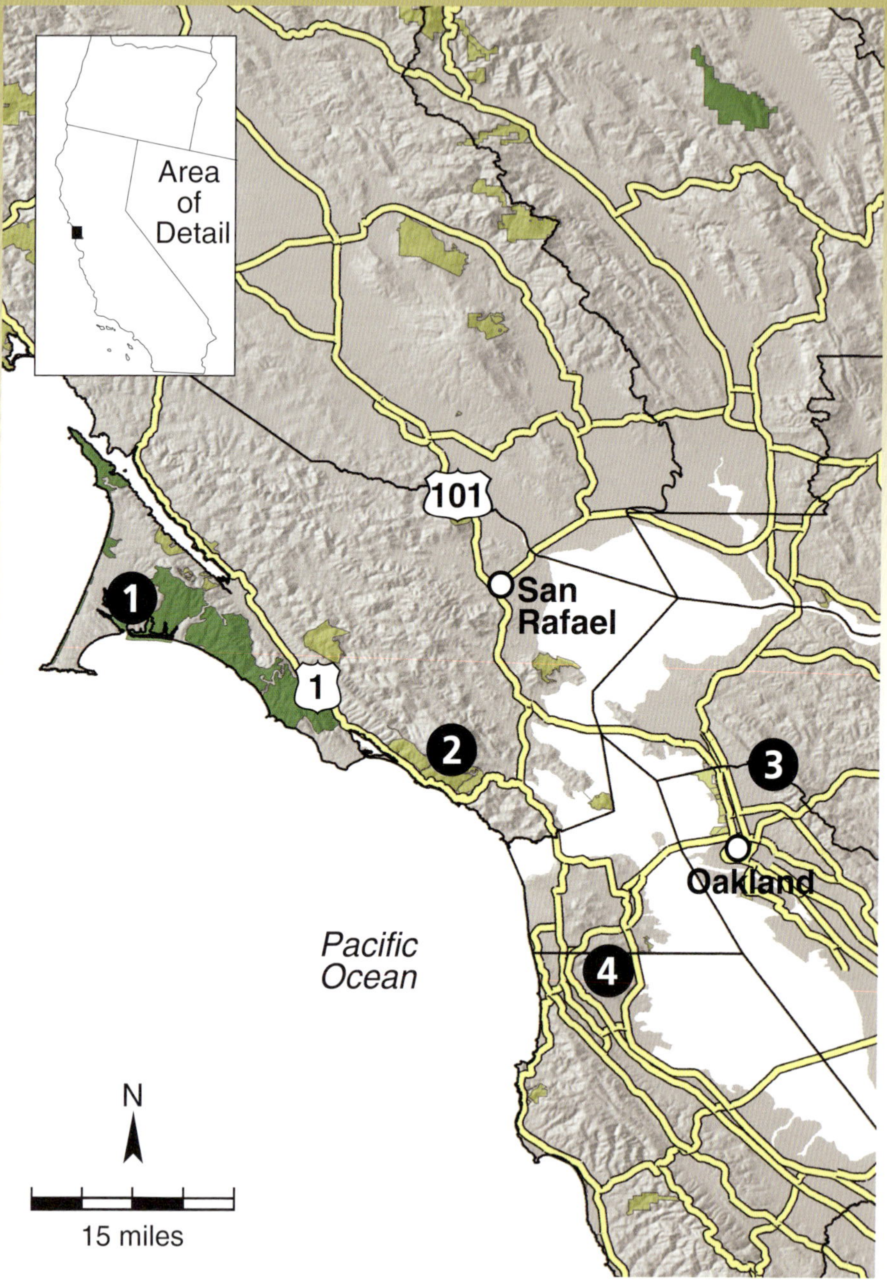

Manzanita Destination: Bay Area

National Forest	State Park
Wilderness	Private Land

Manzanita Destinations: Bay Area

1 Point Reyes National Seashore

Getting there: Travel Highway 1 for 30 miles (50 km) north of San Francisco

Key species: *A. virgata, A. glandulosa* ssp. *cushingiana, A. uva-ursi*

Remarks: The seashore encompasses over 70,000 acres of dunes, sandy and rocky beaches, coastal grasslands, Douglas-fir and Bishop Pine forests, wetlands, chaparral, and lakes. It is a jewel of the California Floristic Province.

2 Mount Tamalpais

Getting there: Take Highway 1 for 24 miles north of San Francisco follow the signs into the park from there.

Key species: *A. montana, A. sensitiva, A. glandulosa* ssp. *glandulosa, A. glandulosa* ssp. *cushingiana, A. canescens*

Remarks: Parks have redwood groves and oak woodlands with a spectacular view from the 2,571-foot peak. On a clear day, visitors can see the Farallon Islands 25 miles out to sea, the Marin County hills, San Francisco and the bay, hills and cities of the East Bay. More than 750 plant species can be found in the park.

3 Huckleberry Botanic Preserve

Getting there: From Highway 24 in Oakland, take the Fish Ranch Road and continue 0.8 miles to Grizzly Peak Blvd. Turn left and go 2.4 miles on Grizzly Peak to Skyline Boulevard, turn left and drive one-half mile

Key species: *A. pallida,* and *A. crustacea* subsp. *crustacea*

Remarks: An ecological jewel, with representatives from a relict plant association found only in certain areas along California's coast where ideal soil and climatic conditions still exist.

4 San Bruno Mountain Co. Park

Getting there: From highway 101, take the Bayshore Boulevard/Brisbane exit and proceed on Bayshore Boulevard to Guadalupe Canyon Parkway, turn west on Guadalupe Canyon Parkway

Key species: *A. imbricata, A. montaraensis, A. pacifica, A. uva-ursi*

Remarks: Parks contains 14 rare plant species and a host of endangered butterflies found in only a few other places in the world.

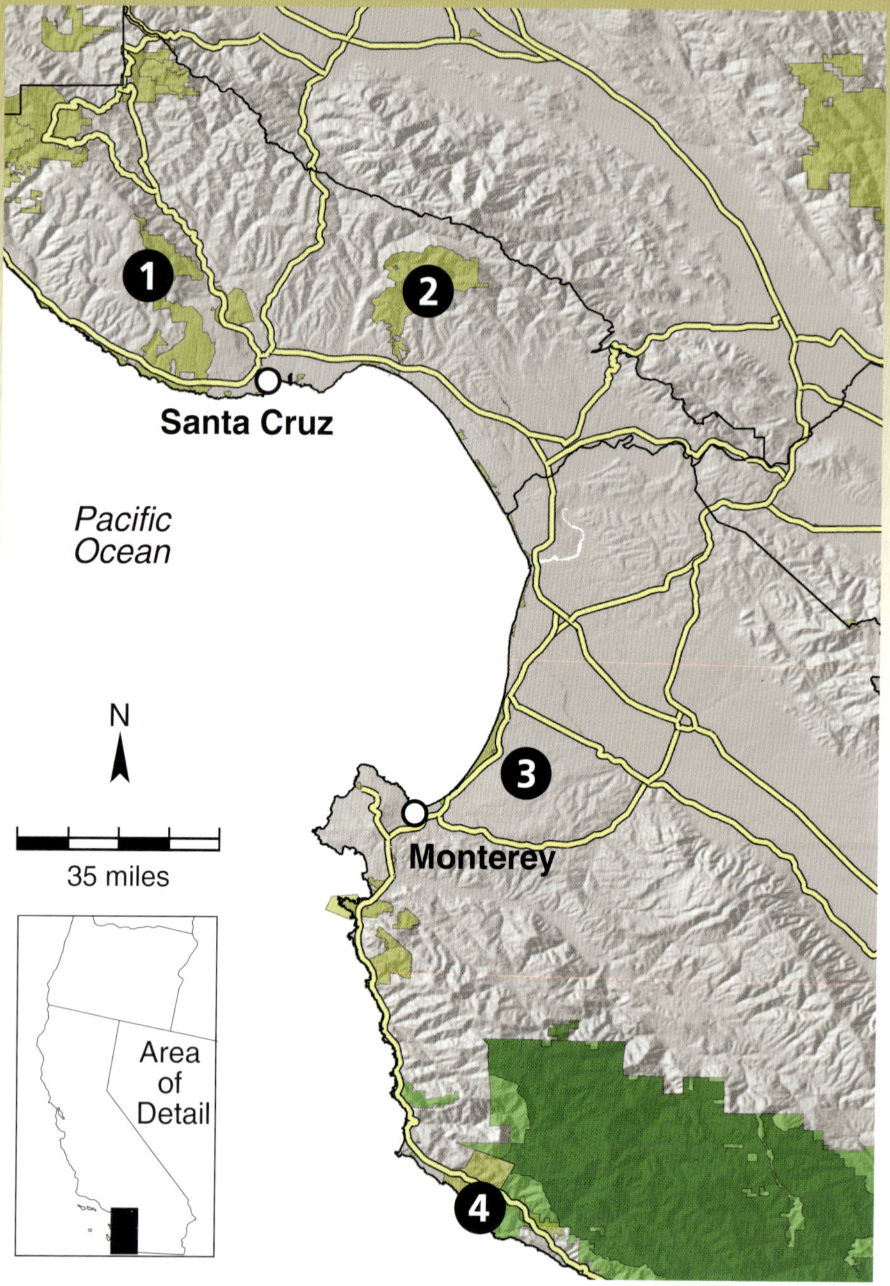

Manzanita Destination: Monterey Bay

Manzanita Destinations: Monterey Bay

1 Empire Grade

Getting there: From Highway 1 in downtown Santa Cruz, take Bay Street to a left on Empire Grade, at the entrance to U. C. Santa Cruz.

Key species: *A. ohloneana, A. andersonii, A. sensitiva, A. glutinosa, A. crustacea* subsp. *crinita* and *A. silvicola*

Remarks: While most of the land along the road is private, there are too many rare manzanita species to not mention it. At the end of the road is the "Lockheed Chalks" which is the property of the Lockheed Martin Corporation.

2 Manzanita Park

Getting there: From Highway 101 in Prunedale, take San Miguel Canyon Rd. north to a left on Castroville Rd. and then onto Manzanita, look for an unpaved pull-off at a corner with trail access.

Key species: *A. pajaroensis, A. hookeri, A. crustacea* subsp. *crustacea*

Remarks: This 450 acre county-run preserve is home to three manzanita species and other rare plants including two orchids and Eastwood's Goldenbush (*Ericameria fasciculata*).

3 Fort Ord National Monument

Getting there: From Highway 101 take exit 326C toward Monterey Penninsula and make a right onto S. Sanborn Rd. Continue onto E. Blanco Road for 1.5 miles, then turn left onto CA-68 W/South Main St.

Key species: *A. pumila, A. tomentosa* subsp. *tomentosa, A. hookeri, A. crustacea* subsp. *crustacea,* and *A. montereyensis*

Remarks: The monument covers over 14,000 acres of a former army facility which was, at one time, home to over 50,000 troops. Now it offers visitors unique manzanita assemblages across its 86 miles of trails.

4 Pfeiffer Beach

Getting there: 30 miles south of Carmel on Highway 1.

Key species: *A. crustacea* subsp. *rosei, A. uva-ursi, A. edmundsii*

Remarks: Julia Pfeiffer Burns State Park features redwood, tan oak, madrone, and chaparral teeming with manzanitas. Trails offer panoramic views of the ocean and miles of rugged coastline.

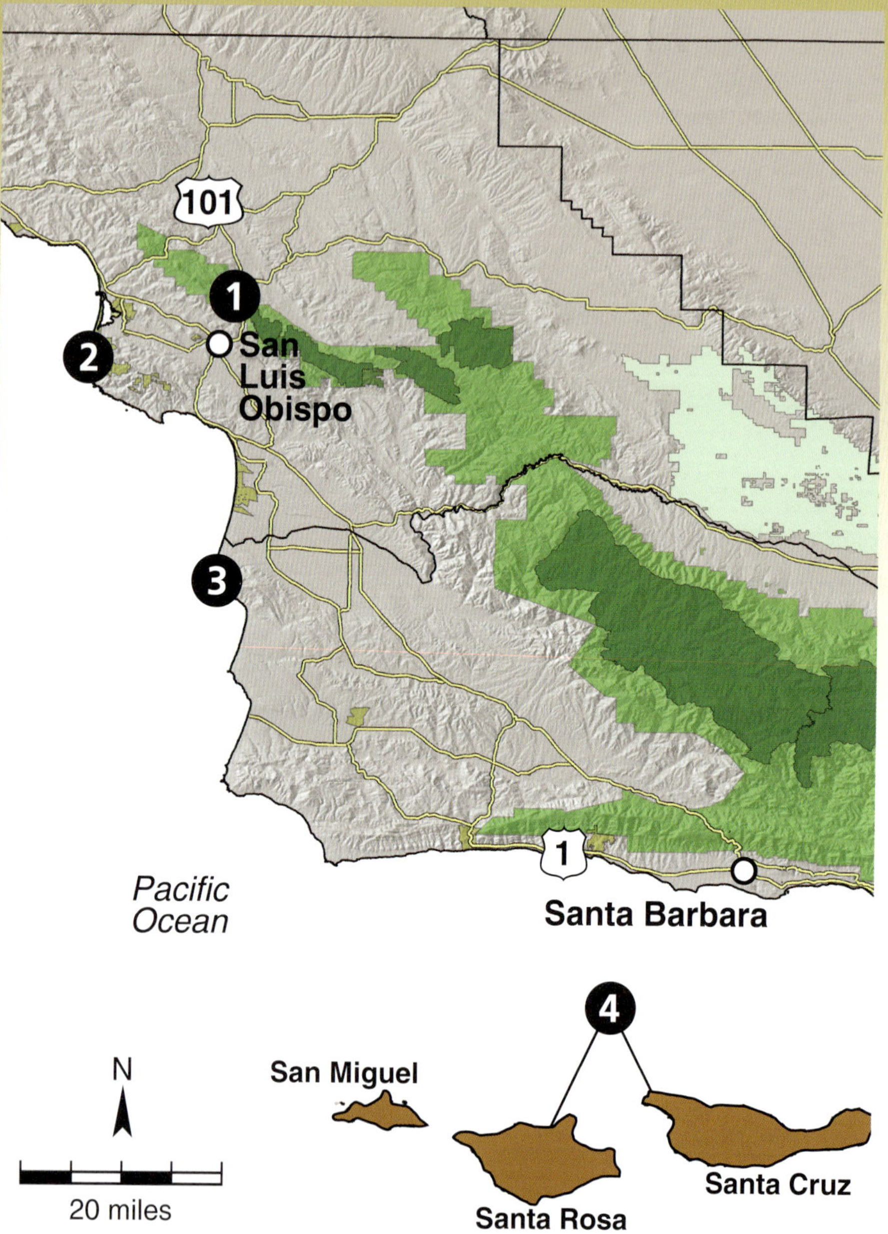

Manzanita Destination: San Luis Obispo-Santa Barbara region

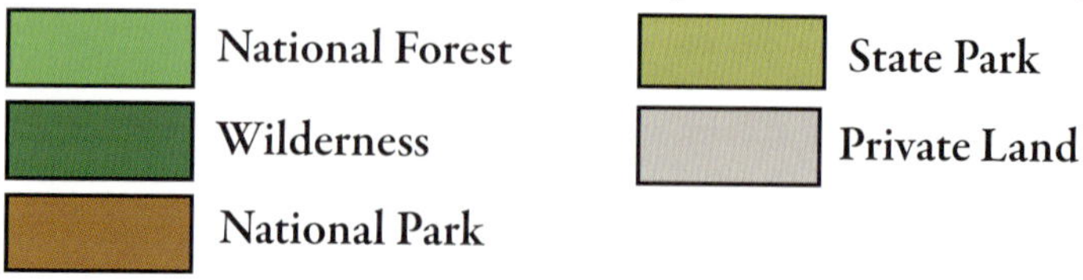

Manzanita Destination: San Luis Obispo-Santa Barbara Region

1 La Cuesta Ridge

Getting there: Follow 101 north from San Luis Obispo.

Key species: *A. luciana, A. glandulosa* (E. Cuesta Grade Rd.); *A. obispoensis* (W. Cuesta Grade Rd.); *A. pilosula* (Pozo Rd @ pass)

Remarks: Grand views overlooking San Luis Obispo to the south, Morro Bay and Morro Rock to the west, the Atascadero hills to the north and the Santa Lucia Wilderness to the east; access trails from Cerro Alto campground.

2 Montaña de Oro S.P.

Getting there: Located at the end of Pecho Road near Los Osos.

Key species: *A. pechoensis, A. crustacea* ssp. *crustacea, A. morroensis*

Remarks: Across over 8,000 acres, plant communities including northern coastal scrub, grasslands, mixed evergreen, riparian, and coastal scub are found here.

3 Point Sal State Beach

Getting there: Located in the north-western part of Santa Barbara County, near the city of Guadalupe.

Key species: *A. purissima, A. rudis, A. crustacea* ssp. *crustacea; A crustacea ssp. eastwoodiana* (SE along Harris Grade Rd.)

Remarks: Park includes 80 acres and just over 1 ½ miles of ocean frontage. There is a 12 mile (round trip) hike to the state beach from Rancho Guadalupe Dunes County Park.

4 Channel Islands National Park

Getting there: Visitor's centers on the mainland in both Ventura and Santa Barbara; islands themselves are only accessible by park concessionaire boats or planes - plan in advance!

Key species: *A. insularis, A. viridissima, A. crustacea* subsp. *subcordata &* subsp. *insulicola*

Remarks: At 96 square miles, its vastness of flora, fauna, and geology makes it a miniature southern California. The island has the greatest number of plant and animal species of all the Channel Islands, including over 60 endemics.

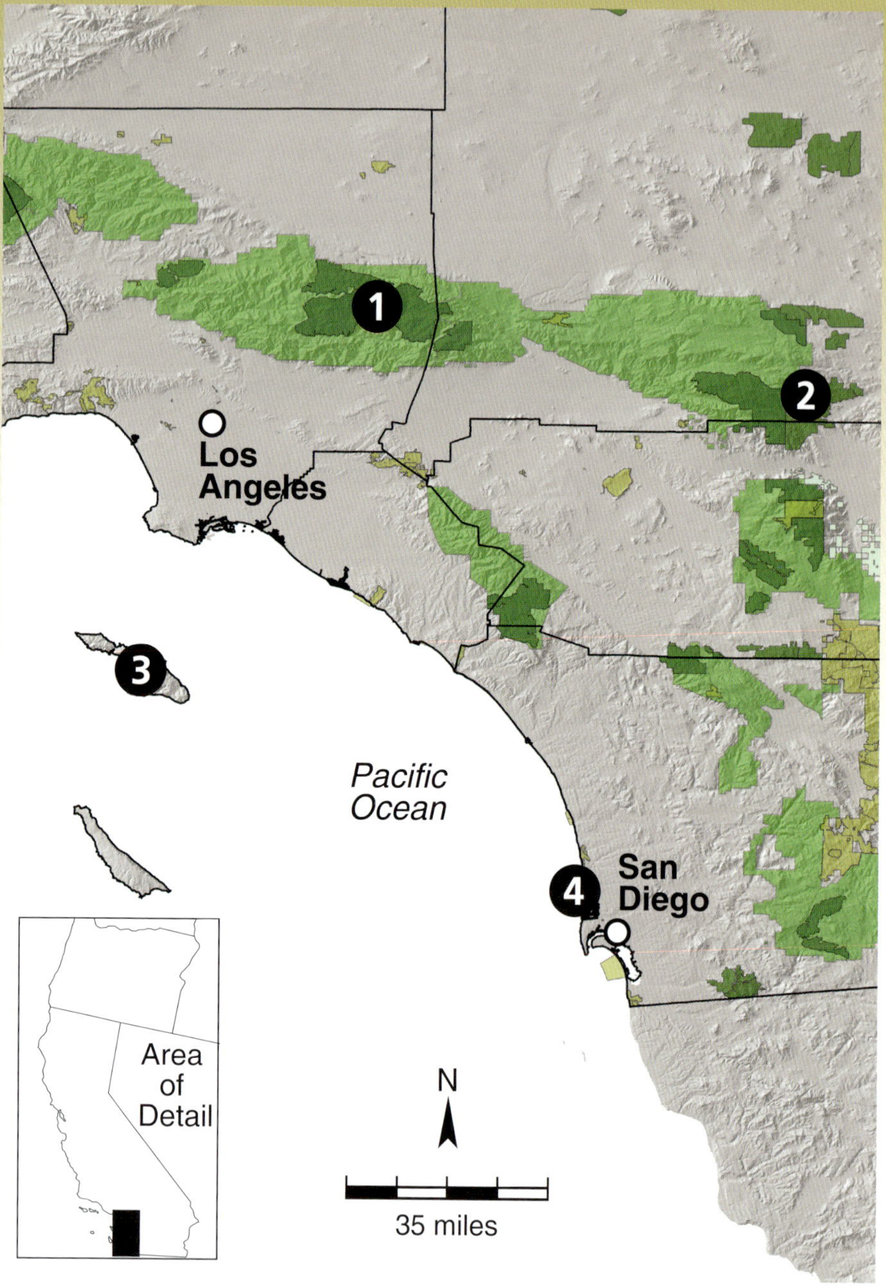

Manzanita Destination: Southern California

National Forest

Wilderness

State Park

Private Land

Manzanita Destinations: Southern California

1 San Gabriel Mountains N. M.

Getting there: Drive the Angeles Crest Highway (Hwy 2) from La Canada to Wrightwood and stop an various trailheads (including the Pacific Crest Trail) along the 66 mile route.

Key species: *A. glandulosa* subsp. *gabrielensis, A. glauca, A. pungens,* and *A. patula*

Remarks: With mountains reaching to over 3000m, the region is a biological island above the urban sprawl of the Los Angeles Basin. Amazing views and unusual ecology can be found in the monument established in 2014.

2 San Bernardino Mountains

Getting there: Barton Flats campground, located 36 miles northeast of San Bernardino, is a good place to start your adventure.

Key species: *A. parryana* ssp. *tumescens, A. pringlei* ssp. *drupacea, A. patula, A. glauca*

Remarks: This area is a hotspot for rare and endemic biota because, like the San Gabriel Mountains, it is a sky island above the surrounding desert. The highest peak is Mount San Gorgonio at 3505m. If you are up for a long hike, this summit offers the best views in southern California.

3 Catalina Island

Getting there: The main option is to take a boat from Long Beach Harbor; take the 710 Freeway south into Long Beach. Stay to the left, follow signs to downtown and exit Golden Shore.

Key species: *Arctostaphylos catalinae*

Remarks: The island ecosystem includes a unique variety of native trees, shrubs, wildflowers, grasses, ferns, mammals, insects, birds, reptiles, amphibians, fungi, lichens, and other forms of life. Some are found only on Santa Catalina Island, while others are widespread.

4 Torrey Pines State Reserve

Getting there: Between La Jolla and Del Mar, California, north of San Diego.

Key species: *Arctostaphylos glandulosa* subsp. *crassifolia*

Remarks: A wilderness island in an urban sea; this fragile environment is the home of our nation's rarest pine tree - the Torrey pine (*Pinus torreyana*) along with other rare plants. Manzanitas grow in the northern extension of park.

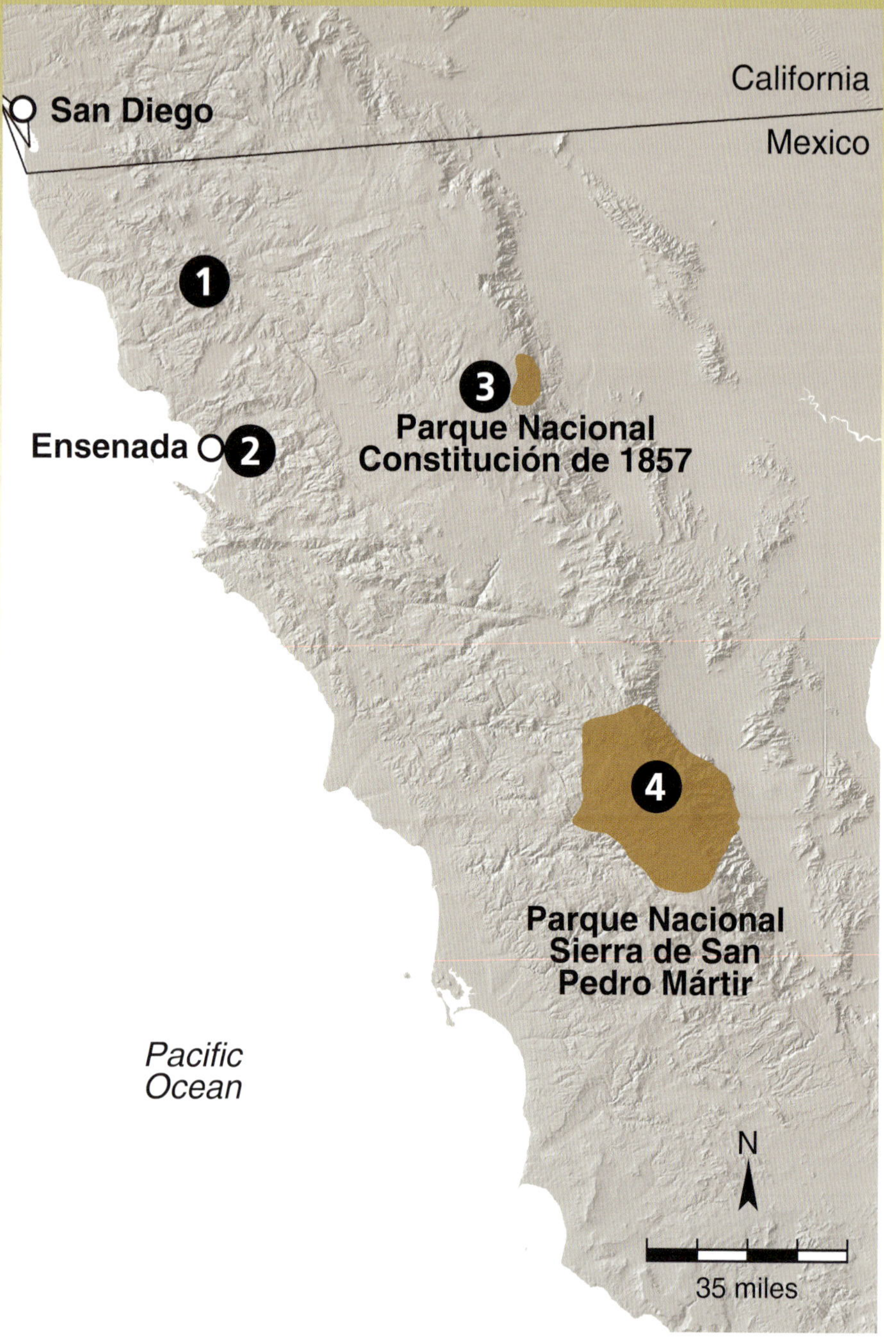

Manzanita Destination: Baja California

National Park

Private Land

Manzanita Destinations: Baja California

1 Cerro Bola

Getting there: Palm Valley is 26 miles south of Tecate, Cerro Bola peak is just to the south.

Key species: *A. bolensis, A. incognito*

Remarks: A few miles south of Valle de la Palma, a steep 4 x 4 road leads to the communications tower at the summit of Cerro Bola. *A. bolensis* is found at lower elevations with *A. glandulosa* ssp. *adamsii* and subsp. *leucophylla* higher. For *A. incognito*, continue south on Hwy 3 for another 10 km to a good dirt road that leads to the village of Carmen Serdán; plants on the south side of road.

2 Ensenada region

Getting there: Ensenada is 90 miles (145km) south of San Diego on Highway 1.

Key species: *A. australis, A. moranii, A. glandulosa* subsp. *atumescens*

Remarks: *A. moranii* on ridges south of village of San Antonia de las Minas along Highway 3. *A. glandulosa* ssp. *atumescens* found west of Maneadero (just south of Ensenada) on road to popular tourist destination, La Bufadora. *A. australis* just east of the town of Eréndira also with *Pinus muricata*.

3 Constitución de 1857

Getting there: Located approximately 60 miles (96km) from Ensenada on federal highway 3, near the village of Ojos Negros.

Key species: *A. peninsularis* subsp. *juarezensis, A. pringlei* subsp. *pringlei, A. pungens*

Remarks: A long dirt road leads to the park which is centered around the lake "Laguna Hansen" a beautiful small lake set among large granite boulders and Jeffrey pines.

4 Sierra de San Pedro Màrtir

Getting there: Follow Mexico 1 for 177 miles (285km) south from San Diego to San Telmo, then travel 48 miles (78km) east on a paved road to the park.

Key species: *A. patula, A. peninsularis* subsp. *peninsularis, A. pringlei* subsp. *pringlei*

Remarks: The southern tip of the California Floristic Province and a place where many species of plants, common to the north, reach their southern range extension. Other rarities include bighorn sheep and California condors.

Manzanita Lists by Region

Arctostaphylos bakeri subsp. sublaevis in The Cedars, Sonoma County

Manzanita species by California county

* Denotes county endemic

Alameda

crustacea subsp. *crustacea*
glandulosa subsp. *glandulosa*
glauca
pallida

Alpine

nevadensis subsp. *nevadensis*
patula subsp. *patula*

Amador

manzanita subsp. *manzanita*
mewukka subsp. *mewukka*
myrtifolia
nevadensis subsp. *nevadensis*
patula subsp. *gankinii*
patula subsp. *patula*
viscida subsp. *viscida*
viscida subsp. *mariposa*

Butte

manzanita subsp. *manzanita*
manzanita subsp. *roofii*
manzanita subsp. *wieslanderi*
mewukka subsp. *mewukka*
mewukka subsp. *truei*
patula subsp. *gankinii*
patula subsp. *patula*
viscida subsp. *viscida*

Calaveras

manzanita subsp. *manzanita*
mewukka subsp. *mewukka*
myrtifolia
nevadensis subsp. *nevadensis*
patula subsp. *gankinii*
patula subsp. *patula*
viscida subsp. *viscida*
viscida subsp. *mariposa*

Colusa

canescens subsp. *canescens*
canescens subsp. *sonomensis*
manzanita subsp. *manzanita*
manzanita subsp. *elegans*
nevadensis subsp. *nevadensis*
patula subsp. *patula*
viscida subsp. *pulchella*

Contra Costa

*auriculata**
crustacea subsp. *crustacea*
glandulosa subsp. *glandulosa*
glauca
manzanita subsp. *manzanita*
manzanita subsp. *laevigata**
pallida

Del Norte

canescens subsp. *canescens*
columbiana
glandulosa subsp. *glandulosa*
glandulosa subsp. *cushingiana*
hispidula
nevadensis subsp. *nevadensis*
nevadensis subsp. *knightii*
*nortensis**
patula subsp. *gankinii*
patula subsp. *patula*
uva-ursi f. *coactillis*
viscida subsp. *pulchella*

El Dorado

manzanita subsp. *manzanita*
mewukka subsp. *mewukka*
nevadensis subsp. *nevadensis*
nissenana
patula subsp. *gankinii*
patula subsp. *patula*

viscida subsp. *viscida*
viscida subsp. *mariposa*

Fresno
glauca
manzanita subsp. *manzanita*
mewukka subsp. *mewukka*
nevadensis subsp. *nevadensis*
patula subsp. *patula*
pungens
viscida subsp. *mariposa*
viscida subsp. *viscida*

Glenn
canescens subsp. *canescens*
canescens subsp. *sonomensis*
glandulosa subsp. *glandulosa*
manzanita subsp. *elegans*
manzanita subsp. *manzanita*
manzanita subsp. *roofii*
nevadensis subsp. *nevadensis*
patula subsp. *gankinii*
patula subsp. *patula*
viscida subsp. *pulchella*

Humboldt
canescens subsp. *canescens*
canescens subsp. *sonomensis*
columbiana
x *media (columbiana* x *uva-ursi)*
glandulosa subsp. *glandulosa*
glandulosa subsp. *cushingiana*
hispidula
manzanita subsp. *manzanita*
nevadensis subsp. *nevadensis*
nevadensis subsp. *knightii*
patula subsp. *gankinii*
patula subsp. *patula*
stanfordiana subsp. *stanfordiana*
uva-ursi f. *coactillis*
viscida subsp. *viscida*

Imperial
glauca

Inyo
nevadensis subsp. *nevadensis*
patula subsp. *gankinii*
patula subsp. *patula*

Kern
glauca
nevadensis subsp. *nevadensis*
parryana subsp. *parryana*
pungens
patula subsp. *patula*
viscida subsp. *mariposa*
viscida subsp. *viscida*

Kings - no *Arctostaphylos*

Lake
canescens subsp. *canescens*
canescens subsp. *sonomensis*
glandulosa subsp. *glandulosa*
glandulosa subsp. *cushingiana*
manzanita
manzanita subsp. *elegans*
nevadensis subsp. *nevadensis*
patula subsp. *patula*
stanfordiana subsp. *stanfordiana*
stanfordiana subsp. *raichei*
viscida subsp. *pulchella* (?)
viscida subsp. *viscida*

Lassen
manzanita subsp. *manzanita*
manzanita subsp. *wieslanderi*
nevadensis subsp. *nevadensis*
patula subsp. *gankinii*
patula subsp. *patula*
viscida subsp. *viscida*

Los Angeles

*catalinae**
glandulosa subsp. *glandulosa*
glandulosa subsp. *cushingiana*
glandulosa subsp. *mollis*
glandulosa subsp. *gabrielensis*
patula subsp. *patula*
parryana subsp. *parryana*
parryana subsp. *tumescens*
patula

Madera

manzanita subsp. *manzanita*
mewukka subsp. *mewukka*
nevadensis subsp. *nevadensis*
patula subsp. *patula*
viscida subsp. *viscida*
viscida subsp. *mariposa*

Marin

canescens subsp. *canescens*
glandulosa subsp. *glandulosa*
glandulosa subsp. *cushingiana*
x *repens (glandulosa* x *uva-ursi)*
montana subsp. *montana**
manzanita subsp. *manzanita*
sensitiva
uva-ursi f. *coactillis*
uva-ursi f. *marinenesis**
*virgata**

Mariposa

manzanita subsp. *manzanita*
mewukka subsp. *mewukka*
nevadensis subsp. *nevadensis*
patula subsp. *patula*
viscida subsp. *mariposa*
viscida subsp. *viscida*

Mendocino

canescens subsp. *canescens*
canescens subsp. *sonomensis*
columbiana
x *media (columbiana* x *uva-ursi)*
glandulosa subsp. *glandulosa*
glandulosa subsp. *cushingiana*
manzanita subsp. *manzanita*
manzanita subsp. *glaucescens*
manzanita subsp. *elegans*
nummularia subsp. *nummularia*
nummularia subsp. *mendocinoensis**
patula subsp. *patula*
stanfordiana subsp. *stanfordiana*
stanfordiana subsp. *raichei*
uva-ursi f. *coactillis*
viscida subsp. *pulchella*

Merced - no *Arctostaphylos*

Modoc

nevadensis subsp. *nevadensis*
patula subsp. *gankinii*
patula subsp. *patula*

Mono

nevadensis subsp. *nevadensis*
patula subsp. *patula*
uva-ursi f. *adenotricha*

Monterey

crustacea subsp. *crinita*
crustacea subsp. *crustacea*
crustacea subsp. *rosei*
*edmundsii**
gabilanensis
glandulosa subsp. *glandulosa*
glandulosa subsp. *cushingiana*
glandulosa subsp. *howellii*
glauca
hookeri subsp. *hookeri*
*hooveri**
*montereyensis**
obispoensis
pajaroensis

*pumila**
pungens
tomentosa subsp. *tomentosa*
tomentosa subsp. *bracteosa**
tomentosa subsp. *hebeclada**
uva-ursi f. *coactillis*

Napa

canescens subsp. *canescens*
canescens subsp. *sonomensis*
crustacea subsp. *crustacea*
glandulosa subsp. *glandulosa*
manzanita subsp. *manzanita*
manzanita subsp. *elegans*
stanfordiana subsp. *stanfordiana*
stanfordiana subsp. *decumbrens*
viscida subsp. *pulchella*

Nevada

mewukka subsp. *mewukka*
nevadensis subsp. *nevadensis*
patula subsp. *gankinii*
patula subsp. *patula*
viscida subsp. *viscida*

Orange

glandulosa subsp. *glandulosa*
glauca
pringlei subsp. *drupacea*

Placer

mewukka subsp. *mewukka*
nevadensis subsp. *nevadensis*
nissenana
patula subsp. *gankinii*
patula subsp. *patula*
viscida subsp. *viscida*

Plumas

mewukka subsp. *mewukka*
mewukka subsp. *truei*
nevadensis subsp. *nevadensis*
patula subsp. *gankinii*
patula subsp. *patula*
viscida subsp. *viscida*

Riverside

glandulosa subsp. *glandulosa*
glandulosa subsp. *adamsii*
glandulosa subsp. *leucophylla*
glauca
parryana subsp. *parryana*
parryana subsp. *deserticum*
parryana subsp. *tumescens*
patula subsp. *patula*
pringlei subsp. *pringlei*
pungens
rainbowensis

Sacramento - no *Arctostaphylos*

San Benito

crustacea subsp. *crustacea*
gabilanensis
glandulosa subsp. *glandulosa*
glauca
pajaroensis
pungens

San Bernadino

glandulosa subsp. *glandulosa*
glandulosa subsp. *cushingiana*
glandulosa subsp. *leucophylla*
glandulosa subsp. *mollis*
glandulosa subsp. *gabrielensis*
glauca
parryana subsp. *parryana*
parryana subsp. *tumescens**
patula subsp. *patula*
pringlei subsp. *drupacea*
pungens

San Diego

glandulosa subsp. *adamsii*
glandulosa subsp. *crassifolia**
glandulosa subsp. *cushingiana*
glandulosa subsp. *glandulosa*
glandulosa subsp. *leucophylla*
glauca
*otayensis**
parryana subsp. *deserticum*
pringlei subsp. *drupacea*
pungens
rainbowensis

San Francisco

*franciscana**
montana subsp. *ravenii**
crustacea subsp. *crustacea*
crustacea subsp. *rosei*

San Joaquin - no *Arctostaphylos*

San Luis Obispo

crustacea subsp. *crustacea*
*cruzensis**
hookeri subsp. *hearstiorum**
glandulosa subsp. *glandulosa*
glandulosa subsp. *cushingiana*
glauca
*luciana**
*morroensis**
obispoensis
*osoensis**
*pechoensis**
*pilosula**
pungens
rudis
tomentosa subsp. *tomentosa*
tomentosa subsp. *daciticola**

San Mateo

andersonii
crustacea subsp. *crustacea*
crustacea subsp. *crinita*
*imbricata**
*montaraensis**
*pacifica**
*regismontana**
sensitiva
uva-ursi f. *coactilis*
uva-ursi f. *leo-breweri**
uva-ursi f. *suborbiculata**

Santa Barbara

*confertiflora**
crustacea subsp. *crustacea*
crustacea subsp. *eastwoodiana**
crustacea subsp. *insulicola**
crustacea subsp. *subcordata**
glandulosa subsp. *gabrielensis*
glandulosa subsp. *glandulosa*
glandulosa subsp. *cushingiana*
glandulosa subsp. *mollis*
glauca
*insularis**
parrayana subsp. *parrayana*
pungens
purissima subsp. *globosa*
purissima subsp. *purissima*
*refugioensis**
rudis
*viridissima**

Santa Clara

andersonii
canescens subsp. *canescens*
canescens subsp. *sonomensis*
crustacea subsp. *crustacea*
glandulosa subsp. *glandulosa*
x *campbellae** (*glandulosa* x *crustacea*)
glauca
manzanita subsp. *manzanita*

Santa Cruz

andersonii
canescens subsp. *canescens*
canescens subsp. *sonomensis*
crustacea subsp. *crustacea*
crustacea subsp. *crinita*
glauca
*glutinosa**
hookeri subsp. *hookeri*
*ohloneana**
sensitiva
*silvicola**

Shasta

canescens subsp. *canescens*
klamathensis
malloryi
manzanita subsp. *elegans*
manzanita subsp. *manzanita*
manzanita subsp. *wieslanderi*
nevadensis subsp. *nevadensis*
patula subsp. *gankinii*
patula subsp. *patula*
viscida

Sierra

nevadensis subsp. *nevadensis*
patula subsp. *gankinii*
patula subsp. *patula*
viscida subsp. *viscida*

Siskiyou

canescens subsp. *canescens*
glandulosa subsp. *glandulosa*
glandulosa subsp. *cushingiana*
klamathensis
mewukka subsp. *mewukka* (?)
nevadensis
patula subsp. *gankinii*
patula subsp. *patula*
viscida subsp. *viscida*

Solano

glandulosa subsp. *glandulosa*
glandulosa subsp. *cushingiana*
manzanita subsp. *manzanita*

Sonoma

bakeri subsp. *bakeri**
bakeri subsp. *sublaevis**
canescens subsp. *canescens*
canescens subsp. *sonomensis*
columbiana
*densiflora**
glandulosa subsp. *glandulosa*
glandulosa subsp. *cushingiana*
hispidula
manzanita subsp. *manzanita*
manzanita subsp. *glaucescens*
manzanita subsp. *elegans*
nummularia subsp. *nummularia*
stanfordiana subsp. *stanfordiana*
stanfordiana subsp. *decumbens*
uva-ursi f. *coactillis*
viscida subsp. *pulchella*

Stanislaus

glauca
manzanita subsp. *manzanita*

Sutter

manzanita subsp. *manzanita*

Tehama

canescens subsp. *canescens*
canescens subsp. *sonomensis*
manzanita subsp. *manzanita*
manzanita subsp. *wieslanderi*
patula subsp. *gankinii*
patula subsp. *patula*
viscida subsp. *viscida*

Trinity

canescens subsp. *canescens*
canscens subsp. *sonomensis*
klamathensis
malloryi
manzanita subsp. *elegans*
manzanita subsp. *manzanita*
manzanita subsp. *roofii*
nevadensis subsp. *nevadensis*
patula subsp. *gankinii*
patula subsp. *patula*
stanfordiana subsp. *stanfordiana*
viscida subsp. *pulchella*

Tulare

manzanita subsp. *manzanita*
mewukka subsp. *mewukka*
nevadensis subsp. *nevadensis*
patula subsp. *gankinii*
patula subsp. *patula*
viscida subsp. *viscida*
viscida subsp. *mariposa*

Tuolumne

manzanita subsp. *manzanita*
mewukka subsp. *mewukka*
nissenana
nevadensis subsp. *nevadensis*
patula subsp. *gankinii*
patula subsp. *patula*
viscida subsp. *mariposa*
viscida subsp. *viscida*

Ventura

glandulosa subsp. *cushingiana*
glandulosa subsp. *mollis*
glauca
parryana subsp. *parryana*
patula subsp. *patula*
pungens

Yuba

mewukka subsp. *mewukka*
mewukka subsp. *truei*
viscida subsp. *viscida*

Yolo

manzanita subsp. *manzanita*

Baja California

*australis**
*bolensis**
glauca
glandulosa subsp. *glandulosa*
glandulosa subsp. *cushingiana*
glandulosa subsp.. *crassifolia*
glandulosa subsp. *adamsii*
glandulosa subsp. *leucophylla*
glandulosa subsp. *erecta**
glandulosa subsp. *atumescens**
*incognita**
*moranii**
otayensis (?)
patula
peninsularis subsp. *peninsularis**
peninsularis subsp. *juarezensis**
pringlei subsp. *pringlei*
pungens

Many of the species common in Southern California continue to grow in northern Baja California, because of the similarity in habitat. Also similar to southern California, enriched species diversity is present in northern Baja, including four species and three subspecies which are endemic.

Other states

Oregon

canescens subsp. *canescens*
canescens subsp. *sonomensis*
columbiana
glandulosa subsp. *glandulosa*
glandulosa subsp. *cushingiana*
hispidula
nevadensis subsp. *nevadensis*
nevadensis subsp. *knightii*
nortensis
patula
uva-ursi
viscida subsp. *viscida*
viscida subsp. *pulchella*

Nevada

nevadensis subsp. *nevadensis*
patula
uva-ursi
pringlei subsp. *pringlei*
pungens

Washington

columbiana
nevadensis subsp. *nevadensis*
patula
uva-ursi

Arizona

patula
pringlei subsp. *pringlei*
pungens
uva-ursi

Utah

patula
pringlei subsp. *pringlei*
pungens
uva-ursi

New Mexico

patula
pungens
uva-ursi

Colorado

patula
uva-ursi

Montana

patula
uva-ursi

Idaho

uva-ursi

Wyoming

uva-ursi

Texas

pungens

Beyond the western United States:

AK, CT, DE, IL IN, IA, MA, ME, MI, MN, NH, NJ, ND, NY, OH, PA, RI, SD, VT, VA, WI.

uva-ursi

Mexico (exc. Baja California)

pungens
pringlei subsp. *pringlei* (?)

Canada

columbiana (British Columbia)
uva-ursi

Guatemala

uva-ursi subsp. *cratericola*

Europe, Asia, Greenland, Iceland

uva-ursi

Glossary

Achene- small, dry, hard, one-seeded, indehiscent (does not split open when ripe) fruit

Abaxial- denoting the lower surface of a leaf

Adaxial- denoting the upper surface of a leaf

Allopatric Speciation- occurs when populations of the same species become isolated from each other, thus preventing genetic interchange

Appressed- pressed closely, but not fused

Auriculate- having one or more structures shaped like an ear or earlobe

Bifacial- leaves different on both sides

Bract- modified, usually reduced, leaf in the inflorescence

Branchlets- a secondary woody stem growing from the trunk or main stem

Bristles- a stiff, strong trichome

Burl- a swelling at the junction of roots and stems that allows sprouting from the base and vegetative regenerate after a fire

Canescent- having a fine grayish-white pubescence

Caudate- tapering into a long, tail-like extension at the apex

Ciliate- fringed on the margin with hairs

Decumbent- lying or growing on the ground but with erect or rising tips

Diploid- one set of chromosomes from each parent

Drupe- fleshy fruit with a thin skin and central stone, or nutlet; manzanitas have what botanists refer to as a *version* of a drupe called a nuculanium

Endemic- state of a species being unique to a defined geographic location

Endocarp- inside layer of the pericarp (or fruit), which directly surrounds the seeds

Exocarp- outermost layer of the fruit wall

Glabrescent (glabrose)- smooth

Glandular/non-glandular- covered (or not) with minute glands

Glaucous- covered with a bloom or smooth, waxy coating

Imbricate- having margins overlapping

Immature- not fully expanded, referring to inflorescence

Inflorescence- complete flower head including stems, stalks, bracts, and flowers

Isofacial- leaves alike on both sides

Lanceolate- lance shaped, much longer than wide

Mesocarp- the middle layer of the pericarp of a fruit, such as the flesh of a peach

Nutlets- a small nut, especially an achene (see also Stones)

Ovate- outline like a longitudinal section of an egg with the basal end broader

Panicle- branched inflorescence with pedicelled flowers

Paniculate- arranged in panicles

Papillate- a minute projection on the surface of a stigma, petal, or leaf

Parapatric Speciation- organisms whose ranges are more or less continuous but are reproductively isolated by the environment, as in soil type

Pedicel- individual flower stalk

Pendant- hanging or drooping forms of organs such as leaves, branches, limbs

Pericarp- the wall of a ripened ovary

Petiole- leaf stalk

Puberulent- a surface covered with very fine downlike hairs

Prostrate - flat, spreading; growing low along the ground

Raceme- inflorescence having stalked flowers arranged singly along an elongated unbranched axis

Scabrous- having a sandpaper-like surface

Setose- thickly set with bristles or bristly hairs

Sessile- stalkless and attached directly at the base

Stones (free/separable)- ability of fruit to be separated or not (see also Nutlet)

Taxa- a group of one or more populations of an organism or organisms seen by taxonomists to form a unit (species, subspecies, variety, etc.)

Tetraploid- two sets of chromosomes from each parent

Tomentose- covered with dense, interwoven trichomes

Trichome- a leaf or stem surface hair

Truncate- cut straight across; ending abruptly almost at right angles to midrib or midvein

Viscid- glutinous; sticky

Suggested Reading

Baldwin, B. G., D. H. Goldman, D. J. Keil, R. Patterson, T. J. Rosatti, and D. H. Wilken, editors. 2012. *The Jepson Manual: Vascular Plants of California*, second edition. University of California Press, Berkeley.

Calflora: Information on California plants for education, research and conservation. 2014. Berkeley, California: The Calflora Database [a non-profit organization]. Avail able: http://www.calflora.org/ accessed: throughout book's creation

Consortium database: Data provided by the participants of the Consortium of California Herbaria. Available: ucjeps.berkeley.edu/consortium/ accessed: throughout book's creation

Gottleib, Leslie. 1968. Hybridization between *Arctostaphylos viscida* and *A. canescens* in Oregon. *Brittonia* 20:83-93.

Harrison, Susan. 2013. *Plant and Animal Endemism in California*. University of California Press. Berkeley, California.

Huang Y, GR Morrison, A Brelsford, J Franklin, DD Jolles, JE Keeley, VT Parker, N Saavedra, AC Sanders, TR Stoughton, GA Wahlert, and A Litt. 2020. Subspecies differentiation in an enigmatic chaparral shrub species. *American Journal of Botany* 107(6): 1–18.

Little, E.L., Jr., 1971. *Atlas of United States trees, volume 1, conifers and important hardwoods.* U.S. Department of Agriculture Miscellaneous Publication 1146, 9 p., 200 maps. Online: < http://climchange.cr.usgs.gov/data/atlas/little/> accessed throughout book's creation.

Jepson, W.L. 1916. "Regeneration in manzanita." *Madrono* 1 (1): 3-11.

Keeley, J.E., M.C. Vasey and V.T. Parker. 2007. "Subspecific variation in the widespread burl-forming *Arctostaphylos glandulosa.*" *Madroño* 54 (1): 42-62.

Marianchild, K. 2014. *Secrets of the Oak Woodlands.* Heyday. Berkeley, California.

Oregon Flora Project. *Oregon Flora Project Plant Atlas. Corvallis: Oregon State University Department of Botany and Plan Pathology.* www.oregonflora.org. accessed: throughout book's creation.

Parker, V.T. 2007. Diversity and Evolution of *Arctostaphylos* and *Ceanothus. Fremontia* 35 (4): 8-11.

Parker, V.T., M.C. Vasey and J.E. Keeley. 2007. Taxonomic revisions in the genus *Arctostaphylos* (Ericaceae). *Madroño* 54 (2): 148-155.

Parker, V.T., M.C. Vasey, and J.E. Keeley. 2009. *Arctostaphylos.* In: Flora of North America, North of Mexico, Volume 8, Magnoliophyta: Paoniaceae to Ericaceae, pp. 406-445. Oxford University Press, New York, Oxford.

Parker, V.T., J.B. Pratt, J.E. Keeley. 2015. *Chaparral.* In: H. A. Mooney and E. Zavaleta (eds.) Ecosystems of California – A Source Book. University of California Press, Berkeley.

Parker, V.T., M.C. Vasey, and J.E. Keeley. 2012. *Arctostaphylos.* In B. G. Baldwin et al. (eds.), The Jepson Manual: Vascular Plants of California, Second Edition, pp. 686-699. Univ. of California Press, Berkeley.

Parker VT, Yunker C, Wechsler G, Vasey MC. 2020. Allopatry, hybridization, and repro ductive isolation in Arctostaphylos (Ericaceae). *American Journal of Botany* 107(12): 1798-1814.

Raven, P. H. and D. I. Axelrod. 1978. *Origin and relationships of the California flora.* University of California Publications in Botany 72:1–134.

Van Dyke, Eric., K. Holl. 2003. *Mapping the Distribution of Maritime Chaparral Species in the Monterey Bay Area.* Prepared for the U. S. Fish and Wildlife Service, Ventura Fish and Wildlife Office

University of Washington Herbarium. Consortium of Pacific Northwest Herbaria. Seattle: University of Washington. http://www.pnwherbaria.org. accessed: throughout book's creation

Vasey, M.C., V.T. Parker, K.D. Holl, M.E. Loik and S.Hiatt. 2014. "Maritime climate influence on chaparral composition and diversity in the coast range of central California." *Ecology and Evolution* 4(18): 3662–3674.

Wahlert, G., V.T. Parker, and M.C. Vasey. 2009. A phylogeny of *Arctostaphylos* (Ericaceae) inferred from ITS sequence data. J. Bot. Res. Inst. Texas

Wells, P. V. 2000. *The manzanitas of California, also of Mexico and the world.* Published by the author.

Visit www.backcountrypress.com/manzanitas for more resources

Manzanita Relatives

The subfamily Arbutoideae contains 6 genera. These genera are found in the northern hemisphere, with most species confined to North America.

- Arbutus (~12 species)
- Arctostaphylos (~60-70 species)
- Arctous (2 species)
- Comarostaphylis (~16 species)
- Ornithostaphylos (1 species)
- Xylococcus (1 species)

Arbutus

- *Arbutus* is the oldest lineage within the subfamily, fossils date back over 50 MYBP.
- *Arbutus* is divided into 2 clades
- One clade is found in western North America including Mexico.
- One clade is found in the Mediterranean Basin.

Arbutus unedo.
Photo by Tom Parker

Arbutus menziesii.
Photo by Tom Parker

Arbutus xalapensis.
Photos by Joey Santore.

Arctous

- *Arctous,* a deciduous genus, is often included in *Arctostaphylos* in many treatments.
- Two species are found in tundra, boreal forest and alpine habitats.

Arctous alpina. Photo by Wolfgang Jauch.

Arctous alpina.
Photo by alpsdake via Wikimedia Commons.

Comarostaphylis

- *Comarostaphylis* is primarily restricted to the Mexican highlands.
- One species, *C. diversifolia,* is found in southern California and northern Baja.

Comarostaphylis polifolia. Photos by Joey Santore.

Comarostaphylis diversifolia on Santa Rosa Island, California. Photo by Michael Kauffmann.

Ornithostaphylos

- *Ornithostaphylos* is a monotypic genus.
- *O. oppositifolia* is characterized by opposite or usually whorled leaves (3 per node).

Ornithostaphylos oppositifolia. Photos by Jeff Bisbee.

Xylococcus

- *Xylococcus* is another monotypic genus.
- *Xylococcus bicolor* is restricted to S. California and northern Baja, originally placed in *Arctostaphylos*; differs in several floral features.

Xylococcus bicolor. Photos by Joey Santore.

Authors & Photographers

Jeff Bisbee

Jeff grew up in the Sierra foothills near Grass Valley, in northern California where he became fascinated with the native plants around his home and began cultivating them in his yard. At age 10, he got his first camera and has been capturing the beauty of nature ever since. He is fluent in Spanish and has traveled extensively throughout Mexico and South America photographing the landscapes and native plants, which have appeared in many publications. He is often accompanied by his wife, Patricia, originally from Buenos Aires, Argentina. Jeff is convinced that the beauty, complex design and adaptability of living things, which produces endless variety, is the work of a wise creator.

Michael Kauffmann

As a youngster, Michael called the eastern deciduous forests of Virginia home. From his bedroom window he would hoot for barred owls and wonder...why are some of those trees still green? He nurtured his evergreen conifer fascination in the Appalachian Mountains during college and, upon graduation, landed a job teaching outdoor education in California. Sans family, alone in the West, he began to hang out with giant sequoias and foxtail pines on weekends. The high country conifers of southern California entertained him for many years until the temperate rainforest eventually called him north to Humboldt County. He now lives in Kneeland, California, with his wife, Allison, and sons Sylas and Asa. Besides writing, he works as a devoted educator and ecologist.

Tom Parker

Tom has been a biology professor at San Francisco State University since 1980. He was raised in Texas, attended the University of Texas, Austin, and discovered California in graduate school at UC Santa Barbara. His first job was at Rider College in New Jersey, but he kept looking for jobs back in California because of the flora and amazing beauty of the state. Manzanitas fell into his life because he was interested in evolutionary ecology and manzanitas have both obligate seeders and resprouting species after fire. Currently, only 9 manzanita species occupy his front yard. He's an active researcher in ecology and evolution, and besides chaparral and wildfire, Tom has investigated tidal wetlands, oak woodlands, mixed evergreen forests, coastal scrub and high elevation Sierran forests.

Michael Vasey

Michael is an ecologist and conservation biologist. A long time lecturer at San Francisco State in the Biology Department, he is also the Director of the San Francisco Bay National Estuarine Research Reserve. Mike has extensive experience as an environmental advocate (formerly on the board of Save San Francisco Bay Association and currently on the board of the Pacifica Land Trust) and a wetland scientist. He also is co-author of the treatment of Arctostaphylos for the second edition of the Jepson Manual and did his dissertation research on the influence of the summer marine layer on chaparral along the central coast of California. He obtained his AB from Dartmouth College, a MA in Ecology and Evolution from San Francisco State, and his Ph.D. from the University of California Santa Cruz.

Nature // Adventure // Stewardship

An independent publisher of web and print media, Backcountry Press explores natural history, ecology, and the western landscape.

Our small business focuses on quality products that enhance the human experience and relationship to the natural world.

Printed responsibly in the United States of America

www.backcountrypress.com

BACKCOUNTRY PRESS
Humboldt County, CA